D0729381

Cultural Geography

How does culture shape the everyday world?

The so-called 'cultural turn' in contemporary geography has brought new ways of thinking about geography and culture, taking cultural geography into exciting new terrain to produce new maps of space and place.

Cultural Geography introduces culture from a geographical perspective, focusing on how cultures work in practice and looking at cultures embedded in real-life situations, as locatable, specific phenomena. Definitions of 'culture' are diverse and complex, and Crang examines a wealth of different cases and approaches to explore the experience of place, the relationships of local and global, culture and economy and the dilemmas of knowledge.

Considering the role of states, empires and nations, corporations, shops and goods, literature, music and film, Crang examines the cultures of consumption and production, how places develop meaning for people, and struggles over defining who belongs in a place.

Cultural Geography presents a concise, up-to-date, interdisciplinary introduction to this lively and complex field. Exploring the diversity and plurality of life in all its variegated richness, drawing on examples from around the world, Crang highlights changes in current societies and the development of a 'pick and mix' relationship to culture.

Mike Crang is a Lecturer in Geography at Durham University.

Routledge Contemporary Human Geography Series

Series Editors:
David Bell and **Stephen Wynn Williams**, Staffordshire University

This new series of 12 texts offers stimulating introductions to the core subdisciplines of human geography. Building between 'traditional' approaches to subdisciplinary studies and contemporary treatments of these same issues, these concise introductions respond particularly to the new demands of modular courses. Uniformly designed, with a focus on student-friendly features, these books will form a coherent series which is up-to-date and reliable.

Forthcoming Titles:

Techniques in Human Geography

Rural Geography

Political Geography

Historical Geography

Theory and Philosophy

Development Geography

Tourism Geography

Transport, Communications & Technology Geography

Routledge Contemporary Human
Geography

Cultural Geography

Mike Crang

Routledge
Taylor & Francis Group

LONDON AND NEW YORK

First published 1998
by Routledge
2 Park Square, Milton Park, Abingdon, Oxon, OX14 4RN

Simultaneously published in the USA and Canada
by Routledge
711 Third Avenue, New York, NY 10017, USA

*Routledge is an imprint of the Taylor & Francis Group,
an informa business*

© 1998 Mike Crang

Typeset in Times and Franklin Gothic by Keystroke, Jacaranda Lodge,
Wolverhampton

All rights reserved. No part of this book may be reprinted or reproduced
or utilised in any form or by any electronic, mechanical, or other means,
now known or hereafter invented, including photocopying and recording,
or in any information storage or retrieval system, without permission in
writing from the publishers.

British Library Cataloguing in Publication Data
A catalogue record for this book is available from the British Library

Library of Congress Cataloging in Publication Data
Crang, Mike.
    Cultural geography / Mike Crang.
      p.   cm. – (Routledge contemporary human geography series)
    Includes bibliographical references and index.
    1. Human geography. I. Title. II. Series.
  GF43.C73   1998
  304.2–dc21   97–45070

ISBN 0–415–14082–X (hbk)
      0–415–14083–8 (pbk)

Contents

List of figures

List of boxes

Acknowledgements

Putting this textbook together has been the result of many conversations, questions and encounters both with colleagues and students. I should first say that my prime guides have been tutees here at Durham and this book is in large part a response to their questions, problems and comments. I should also thank Emma Mawdsley and Peter Atkins for reading through drafts and pointing out the unclear passages that were merely hiding behind poor phraseology. David Bell as editor of the series has been with this book from the start and I should thank him and Sarah Lloyd at Routledge for bearing with its hesitant progress.

Permission to reproduce pictures here is gratefully acknowledged from the following:

Fine A.C.
Bibliothèque Nationale, Paris
Ethical Consumer Magazine
J. Sainsbury plc
Morocco Tourist Board
Teesside Development Corporation

Every effort has been made to contact copyright holders and we apologise for any inadvertent omissions. If any acknowledgement is missing it would be appreciated if contact could be made care of the publishers so that this can be rectified in any future edition.

Locating culture

- **What do we mean by culture?**
- **Why is it studied?**
- **What sort of things will it involve?**

It seems obvious that a book introducing students to cultural geography must start with a definition of what it is about. Obvious, but almost unfeasibly difficult. Defining the word *culture* is a complex and difficult task which has produced a range of very different definitions. In some ways 'cultural geography' is easier to grasp than simply trying to define either of its component parts. This is because, despite occasionally sounding the most airy of concepts, this book will argue that 'culture', however defined, can only be approached as embedded in real-life situations, in temporally and spatially specific ways. This book focuses on how cultures work in practice. The philosophy of this book is that this is the contribution of geography – insisting on looking at cultures (plural) as locatable, specific phenomena.

There seem to be two typical reactions to the idea of cultural geography by new students. The first is to think of the different cultures around the globe, to think of the sort of peoples presented in documentaries such as *Disappearing World*. In this vision, cultural geography studies the location and spatial variation of cultures; it is a vision of peoples and tribes echoed in *National Geographic* magazines and travel stories. The second reaction is to associate culture with the arts, with 'high culture', that is, and is normally followed by a slightly perplexed look as to what geography can have to do with that. Both versions capture only a tiny part of what is dealt with as 'cultural geography'. It has been one of the fastest expanding, and, in my admittedly partisan view, one of the most

Box 1.1

Defining culture

By the 1950s authors could collect over 150 different definitions of culture in use in academic books. This book is not trying to push a very specific definition. Indeed the different approaches recounted here may imply rather different ideas of what culture is. The guiding principle in this work is that cultures are sets of beliefs or values that give meaning to ways of life and produce (and are reproduced through) material and symbolic forms. In this way I want to avoid two notions especially. The first is a depiction of culture as a sort of 'residual variable' for all those things not accounted for in other fields, the 'remainder' they don't explain. I argue culture is far more central than such accounts allow. Second, the mention of 'way of life' questions how much the individual can pick and choose, while the reproduction of them brings in issues of change over time. The possibility that current societies may actually develop a more 'pick and mix' relationship to culture is something that is developed through the book.

interesting sub-disciplines in geography over the last fifteen years. The reason is that its subject matter is so wide-ranging both in location, issues raised and the type of material involved. Let me try to show why this is so by starting to locate what cultural geography involves.

Travellers' tales

The initial assumption of many is that cultural geography is about how different cultures live in different parts of the globe. One of the prime motivations for many students doing geography is a fascination with the diversity of human life. Undoubtedly the diversity of people on the globe is an important starting point but one that needs to be developed further. Different groups are marked out not only by different clothes, ornaments, lifestyles but are also guided by different 'world-views', different priorities, different belief systems, different ways of making sense of the world. Cultural geography thus looks both at the forms of difference, the material culture, of groups but also at the ideas that hold them together, that make them coherent. This means this book will look not only at how cultures are spread over space but also at how cultures make sense of space. This book will thus track the ideas, practices and objects that together form cultures – and how these cultures form identities through which people recognise themselves and others. It will track through a range of scales as it ponders the role of states, empires and nations, firms and corporations, shops and goods, books and films in creating identities. Cultural geography looks at the way different processes come together in particular places and how those places develop meanings for people. Sometimes we may be looking at processes of a global scale, at other times we might be interested in the *micro-geography* of houses, the intimate and personal scale of things that form people's everyday world.

So cultural geography is about the diversity and plurality of life in all its variegated richness; about how the world, spaces and places are interpreted and used by people; and how those places then help to perpetuate that culture. This book will thus have to deal with how ideas and material, practices and places, cultures and space interrelate. You will find no single answer; rather, the chapters show different cases and different approaches people have taken to these issues.

Cultures are not just about exotic faraway peoples, but also about the way we, in the West, do things. It is all too easy to take your own culture as in some sense natural and look at the peculiarities of other groups. As Pierre Bourdieu puts it, any culture is a riot of colours and discordant sounds until you learn the rules that guide and make sense of it (Bourdieu 1984). The implication is that every culture has rules that are as arbitrary and surprising as every other. Thus the anthropologist Marshall Sahlins once famously remarked, we call India the land of the sacred cow because some Hindu customs seem strange – this animal is allowed to wander where it wishes, despite being edible it is not eaten, and it defecates wherever it goes. Of course, Marshall Sahlins points out, by the same criteria we could marvel at the UK or the USA as the land of the sacred dog (Sahlins 1976). In this sense cultural geography tends towards a relativistic stance. We must acknowledge the particularity of our own culture and not sit in hasty judgement over other cultures.

gh culture, popular culture and the everyday

If we look around at any society, there are activities whose primary role is symbolic; say, theatre, opera, art, literature or poetry. All of these are generally seen as products or expressions of that society's culture. Indeed we might instantly extend our account to include the libraries, museums, galleries and so forth that allow these forms to exist; that preserve and reproduce them; that make them available to people. Cultural geography must thus include the institutions that keep cultures going. This in itself might take us into surprising corners – into, say, schools where children are taught the 'great' figures of their culture's history or literature, or maybe the interpretation of different public monuments. But even if we stick to symbolic things we would have to conclude that modern society is just as full of rituals and ceremonies as ones distant or remote from us. British people might have to include the rituals of royalty (the opening of Parliament, the Trooping of the Colours), in the USA the 4th of July, or Bastille Day in France. All these are festivals or rituals sanctioned or promoted by the state, so already cultural geographers might be asking why the state promotes certain rituals, and not others, and what it gets out of them. But culture extends further than just state-sponsored rituals. There are vast numbers of different festivals and rituals supported by different religions and the cultures associated with them; we would have to admit Christmas, Thanksgiving, Passover and Ramadan among many more, as festivals that sustain and reproduce different cultures.

Nor can we stop at religion: culture spreads further into our lives and societies. We might say that Christmas is a religious festival and draws on a Christian culture – but equally for most people it is a family festival that draws on consumer culture. So the role of manufactured goods, mass consumption and so forth would form part of some studies. We would have to say Valentine's Day or Hallowe'en bear even less connection to a sponsoring religion – we have secular and often commercial festivals. We would have to include popular festivals such as Guy Fawkes Night in Britain or Burns' Night in Scotland. However, culture is not confined to festivals and holidays; it pervades everyday life. So we might also include folk culture – looking at indigenous dialects, vernacular architecture and so on. But in the contemporary West the same logic means we need to include 'ordinary' culture, not just historic elements but the everyday. That means we need to think how people assemble meaningful worlds out of mass-produced goods, how they relate to places through films and books, how such cultures are bound up in work and leisure.

One of the things this book will try to illustrate is that cultures are often political and contested – that is, they mean different things to different people in different places. So the state may promote one vision of a 'people' through particular symbolic locations. Other groups might offer alternative symbolic geographies or might attribute very different meanings to the same places. In this way this book will look at how power and meaning are written on to the landscape. How monuments and buildings may be used to try to bind people together, to stress common interests, to promote group solidarity.

One of the clearest ways in which different cultures have reproduced themselves is through territorial segregation. Such a process can be seen in cities where different gangs mark their territory in graffiti. On a less intense level we can see the same process in supporting different football teams. Whether there are religious connotations such as Celtic versus Rangers, the wearing of team colours and so on suggests this is another smaller set of cultures. The contemporary city may house a vast range of peoples rubbing shoulders, buying different foods, creating festivals and music, forming a dense and variegated cultural mosaic. How these cultures relate over distance, how they bring formerly distant identities and put them together forms a fascinating element of contemporary cultural geographies. Cultural geography must then look at the fragmented juxtaposition of cultural forms and the identities arising from this. We thus need to consider how cities and nations may contain a plurality of cultures. We might term these *sub-cultures*. We need only think of the different worlds created through Rave culture and clubbing. These spaces offer a different social milieu, a different set of practices, sustained by a very different geography of venues than that associated with 'official' British culture. We could add to this by looking at how the gay community sustains itself and fosters a sense of identity through the spaces and practices of clubs and shops.

All this indicates we need to see both how particular sites acquire meanings and how places and sites are used by cultures. Let me illustrate this by looking briefly at student culture and how much of this is embedded in particular spaces and geographies. First, it is a geography of bringing people together; second, and symmetrically of separating these people from both the resources and constraints of the areas they come from. For the average fresher in the UK, there is a sudden pressure of new people, new rules of the game, a setting free from strictures of a parental home, yet also a loss of the support that home provided. A process that is supported around a geography of places – the bars and 'student-friendly' pubs where students can meet new people, the halls of residence, the canteens and faculties around which networks of acquaintances can be formed. The student community is stitched together out of these places; it relies on this geography. The provision of single study-bedrooms creates a private space which students control and can personalise, which they can invite people to or retreat into. Places gain meanings, lecture halls obviously are about learning (and possibly sleeping), libraries also are about learning but both also are places to meet people. Students may grow attached to 'their' department or faculty, maybe enjoy the symbolism of great halls or whatever on graduation. What this suggests is the way particular spaces and geographies are deeply involved in maintaining cultures all the time. And these cultures are not just about overt symbolism, but about the way people live their lives. The example above shows that the material things enabling students to live work and play are all involved in maintaining a culture.

Economy and culture

It should be clear that the separation of the economy and culture is problematic. In fact it is possibly a hallmark of modern capitalist cultures that they treat the economy as in some way separate from the rest of the culture. But if the two are to be analysed separately how should the relationship between them be seen? The most influential approach, from a number of perspectives, is to look at culture as some sort of clothing, superstructure, barrier to rationality or remainder after the economy has been dealt with. We shall come across these again but I shall introduce them here to warn the reader not to use these as implicit models. The first two models see culture as providing the symbolic face behind which the 'real' economy works. In early Marxist accounts the economy determined social relations that were reflected in particular cultural forms. In the other approaches, culture is treated as that which an economic analysis cannot explain. Thus geographers (and economists) deploy questionnaires to look at optimum-location decisions and, because these do not fully account for location, 'personal preferences' or cultural factors are introduced as a sort of remainder once the economic is accounted for. Likewise, in accounts of indigenous farmers' reactions to agricultural techniques imported from the West, their local culture is portrayed as 'local', peculiar and a barrier to accepting Western progress.

The primacy of 'economic' explanations has to be questioned since it is very easy to reverse the normal accounts. Thus instead of the economy determining cultures we can reverse that. To use Sahlins again, he points out the enormous amount of economic activity that is structured around men wearing trousers and suits and women skirts and dresses (1974). Think, he asks, of the consequences for hundreds of factories if that changed. Likewise we could return to look at food, and trace how changing tastes for food have altered economic systems over and over. Think how much of the Caribbean economy is based around a Western taste for sugar, or how much of India's has been linked to the taste for tea. Of course thinking in this way does not change the separation of culture and economy it just inverts the relationship. This book will argue that we need to avoid seeing either culture or the economy as determining the other; and indeed in many cases that it is more helpful to see how they interact than to separate them. This will be a central part of Chapters 8 and 9.

sitioning culture

So far then I have tried to establish that culture cannot be locked away in either distant peoples or in high art. Culture is part of our everyday lives, indeed it is what gives meaning to those lives. I have equally tried to stress how cultures can be seen to change, to be contested. Finally I have tried to show how these cultures are reproduced through a range of forms and practices embedded in spaces. How we might then approach cultures and these spaces is a complex issue. Essentially the rest of this book begins to see how different approaches have developed around these issues. I do not suggest that one approach can be isolated as reigning supreme. They tend to be looking at rather different situations. Chapter 2 opens with a 'traditional' approach to space and culture. It attempts to use 'material culture', artefacts and products, to see how different cultures inhabited different areas creating distinctive *cultural landscapes*. Chapter 3 looks more closely at how landscapes may be intentionally shaped by people to carry meanings – looking at the *iconography* of places. Landscapes are not solely interpreted through direct contact so Chapter 4 explores what we might call *literary landscapes* – the geographies created by books and novels. This chapter thus looks at the relationship of books to places, of how places may be affected by popular books and also how space is used within books to create a textual landscape. Following this, Chapter 5 looks again at the relationship of space and literature, but in this case looks at how popular literature deals with cultural difference. The focus of this chapter is on literature of the imperial period and it looks at how it helped shape Western views of cultures around the globe. Chapter 6 takes many of the same approaches and relates them to film and music, looking for continuations and differences in these different media. Chapter 7 takes a slightly different focus on global issues and raises questions about how people relate to places in a globalising world. It introduces ideas of a *humanist* geography, where personal meaning is the crucial category for geography and where a *sense of place* and attachment to place may be imperilled in a *placeless* world. Chapter 8 takes issue with many of the fears in the previous chapter; it focuses on how people construct meanings through mass-consumed goods, and how they may not be a threat to meaningful places, as it looks at different geographies and spaces of consumption. Both these chapters thus deal with how places can be constructed in ways that deliberately evoke faraway cultures that mediate cultural difference. Chapter 9 turns its attention to spaces of production looking at how different forms of work produce or utilise different cultures of accepted behaviour. Examples are taken from global industries, and the service sector where workers' cultures can be a part of the product. Chapter 10 picks up the themes of global change and cultural difference to pose questions about how culture is used as a basis for nationalism. In the contemporary world it makes an appeal to move away from the association of one culture with an area and rather look at the *hybrid* forms springing up as cultures meet. The concluding chapter asks questions about the role of the cultural geographer in all this, suggesting that we should see science and academia as another culture rather than independent of what is being studied.

Mapping cultural geography

If we have thus far thought about what sort of things cultural geography might study, we need to be aware that this has evolved over time. The chapters thus reflect, in part, the changing dynamics of the sub-discipline. We might call this the historical geography of cultural geography. The metaphor of an intellectual landscape can be helpful. If we think of a disciplinary 'map' we would see the blurred borders and traffic between areas of interest, we would see the traffic and paths leading out into other disciplines. Over time we would see shifting centres of population, shifting cores and peripheral topics. The landscape would be far from static. But we should hesitate before going further with this exercise with two cautionary thoughts. First, the evolution of cultural geography is bound up in wider disciplinary changes and changes in the social sciences and humanities, and also, at a wider scale, with shifts in society in general. This map is thus one tiny fragment that really needs embedding in a larger picture. Second, through the course of this book one of the key points will be that interpretations are linked to specific points of view and are contestable. Charting disciplinary history is no exception to this rule. What follows must be a necessarily partial view. Perhaps the mapping metaphor is unfortunate, since it suggests some overview of the scene as though we could float free of all baggage and find some perfect viewpoint to see the true outline of events. Floating free like this is impossible. As I write this, I cannot pretend to have either a perfect knowledge or to be detached from the events I talk about. If I was, why would I be spending my time thinking about them? Even with the best will in the world, I cannot give an absolute account. Rather than floating on high we might better think of this as constructing some field sketches as we go by, trying to work out how things fit together. The next corner, the next turn might make us change our ideas all over. Though some may find this a dispiriting thought, this not getting the 'answer', it is, on the contrary, one of the things that makes study in this field so exciting: the field is not dead and fixed, but continually changing.

We might locate the original sparks of cultural geography back in the sixteenth century in the ethnographies of such as Lafitau or Lery, describing the peoples and customs of the New World. We might look at the literary and allegorical fields opened up at the same time by writers such as Rabelais or later by Swift, which used imagined and real travels to map out the cultures of their own societies. The connection of these real and imagined places, and the role of the strange or exotic is something that is now being re-examined and represents an intersection of geography and anthropology from the earliest times. It also connects both disciplines to the European imperialist project with all the problems that leaves for them. It also draws us towards two more commonly thought of landmarks. The concern with race and imperial development marks the work of the German political theorist Ratzel's *Anthropogeographie* from the end of the nineteenth century. He used a metaphor – imported from the burgeoning field of Darwinian biology – to suggest we should treat cultures like organisms. He identified cultures with *Volk*, or peoples, defined on ethnic and cultural differences. As in Darwin, he saw a struggle to prosper and survive between these cultures and mapped this out territorially as struggle for *Lebensraum* ('living space'). Vibrant cultures would expand and come to dominate or displace less 'vital' ones. The connections of this to the projects of imperial expansion and its later borrowing by Nazi ideology form a bleak reminder in our landscape. A school of thought related to this was developed largely in America, especially around Ellen Semple in the first quarter of the twentieth century, known as environmental determinism, which took the territorial units of Ratzel and linked them to basically climatic conditions. This school looked at how cultures evolved in response to the natural environment through adaptive behaviour (again borrowing the central metaphor from biology). This was, however, not the most powerful influence on cultural geography in the USA. Chapter 2 takes up the story of how these ideas were challenged by Carl Sauer and what came to be the Berkeley school of cultural geography. He suggested a much more nuanced relationship of people and environment not one-way causation or simple biological analogies. Teaching in Berkeley until the 1970s he had a vast influence on cultural geography in the USA. It developed connections with biogeography and material anthropology with a focus on the material culture of peoples. Picking up on the encounter with the New World it led to studies of how people shaped and reshaped the landscape, how cultures travelled and changed, how immigrant peoples set about reshaping the landscape of the Americas and the artefacts that embodied their efforts.

This legacy has come to be the cause of some friction between US and UK geography. First, the rural and historic bias of the work did not reflect urban life and experience. Thus geographers during the 1970s and 1980s such as David Ley or Peter Jackson looked for inspiration instead to the work of urban sociologists such as those of the 'Chicago school'. The latter had reacted to the 'melting pot' of a city that doubled its population in twenty years and brought together people from every part of the US and Europe. In schools their work has been picked out in terms of Burgess's concentric-ring model of the city. This is something of a travesty of the work done by Burgess, Park and others. The bulk of their work was involved in studying the 'urban villages' and sub-cultures that were forming in the city – from Little Italy to Skid Row. From their work cultural geography drew in ideas about the fragmentation of cultures as 'ways of life' and a methodology of 'ethnography', studying people by living among them. A second conflict thus sprang up about how culture itself was seen – with arguments that the Berkeley school still had the 'organic' metaphor found in Ratzel. These issues will be expanded on in Chapter 2. Suffice to say, in British geography especially but also in the US, a series of new approaches developed which looked to the symbolism of cultures.

At the same time, a concern for individuals and their experience was marked by a humanistic school in geography (Chapter 7) which sought to talk not of peoples and folk, but of actual people and ordinary folk as individuals through their lived experiences. It connected with philosophical ideas (called phenomenology, see Chapter 7) and

reinvigorated the idea that geography was an interpretative art. This
emerged in part as a response to the rise of quantitative and systematic
approaches in geography from the 1960s. Forming something of a fusion
of these two strands of thought there emerged perhaps the first new
cultural geography drawing on psychology and in particular ideas of
behaviourism. This developed into work on decision-making, on to the
mental maps people held of the city or world – in short the geography
inside people's heads. Of course this rapidly became a criticism that too
much was about individuals, not collective cultures, and too much was
about ideas not the material world. The relationship of this internal
world-view with the outside world became a major field of inquiry on
both sides of the Atlantic. Emerging from this were studies on how
people related to landscape, looking at perceptual processes as well as
material and aesthetic interpretations. Picking up the Berkeley interest in
the material landscape we find this coupled with a more interpretive
understanding of everyday places. This was later adapted and
transformed with studies of the shaping of symbolism in landscape and
its representations (see Chapters 3 and 6).

Geography and the social sciences were also changing sharply in the
light of decolonisation, the Vietnam War and the rise of Marxist theories.
In geography these swept through mostly economic geography which
became possibly the centre of human geography through into the 1980s
as various ideas of political economic interpretation developed. By the
late 1980s though cultural geography in the UK at least was assuming a
new and probably unexpected centrality – with a so-called 'cultural turn'
spreading out to reformulate not only cultural geography but other sub-
disciplines. A reformulated cultural geography was taking up ideas of
Marx and humanism in looking at the struggles and contests over
interpreting cultures. Alongside these ideas the whole of the human
sciences were having to consider a 'post-colonial' critique which asked
questions about how much of conventional thinking was still in thrall to
ideas that dominated during imperialism – whether such ideas were
Eurocentric or fatally flawed. Fundamentally it questioned some of the
most stable landmarks of the intellectual landscape and asked whether
they fitted in a new pluralistic world. From another direction, by and
large French or Continental philosophy (as opposed to Anglophone
schools), came a post-structuralist critique of how models of society
operated. The ideas of rational and scientific, reductionist accounts of
society came under intense scrutiny as did the grand stories of social
evolution and economic development. If the intellectual landscape was

thus shifting, in society at large the categories of class and labour mobilisation seemed to be being supplanted by identity politics. Movements for women's rights, gay rights, civil liberties and indigenous peoples were using ideas of shared identity or sub-cultures. In the UK this inspired the rise of a cultural studies informed by the work of the Chicago school, in the US a cultural studies that was perhaps a little more literary in outlook.

These shifts to paying more attention to people's identities came late to geography, but came with a vengeance. Part of these shifts were summed up in heated debates over postmodernism which criticised the assumptions of conventional geography (the sort that lie behind Peter Haggett's much-reprinted book *Geography: A Modern Synthesis*, for example. Post modern approaches suggested the assumptions of the 'modern' meant that such a synthesis excluded as much as it brought together). Another part of the postmodern challenge to conventional academic work was in heated debates within feminism over the politics of knowledge; another was a challenge to development studies; yet another was a challenge to the narratives of Marxism from within and without. In each case questions were raised about whose rules were being used and whose identities were taken as normal and who was excluded or overlooked. The result was that culture, from being a rather neglected afterthought in most studies, became a central issue.

Summary

This chapter has suggested that cultural geography must look at things beyond high culture, at lifestyles in the West as much as in remote peoples and at the way spaces are used as well as the distribution of peoples over space. It has suggested that the separation of economic and cultural is often problematic and that it led to a misguided privileging of the economic in much geographical work. The chapter then briefly outlined the development of ideas in cultural geography before suggesting how these would be worked through in the rest of the book.

Further reading

It will be apparent that different themes running through this book will link to other books in this series of textbooks. Thus chapters 8 and 9 will connect with issues discussed in the Economic Geography book in this series. The chapter on symbolic landscapes will find echoes in the Historical Geography book. Issues raised about how we interpret cultures in the latter half of this book and especially the conclusion will resonate with ideas in the Theory in Geography book. Ideas in Chapter 5 around imperialist literature will also serve as a background or prelude informing current debates over development studies. Here it is worth suggesting that if you wish to look more generally at issues in cultural geography you might look at journals such as *Ecumene* or *Society and Space* and see what sort of topics are covered there. Relevant material will also appear in non-geography journals, so if you wish to follow up ideas around film you might look at a journal such as *Screen* or for television, *Media, Culture and Society*. These sorts of journals will contain papers at the leading edge of ideas so they may be hard to follow at first. More specific ideas of further reading will follow each chapter suggesting key writers or works for particular topics.

2 People, landscapes and time

The region is a medal struck in the likeness of its people

Paul Vidal de la Blache

- Cultural landscapes, culture areas and the 'regional personality'
- Material culture, artefacts and the landscapes
- Distribution, diffusion of cultural forms

Anyone looking around the world can see a vast mosaic of different peoples with different customs and beliefs. This has been the starting point of a whole tradition of cultural geographies, concerned with the landscapes created by different groups in different places. This chapter will outline some of these approaches. In particular it will highlight the *material culture* of these groups as a process of transforming the environment. It will do this by following the work of the Berkeley school and looking at the commonalities it shares with the *Annáles* school in France and the local history approaches in the UK. In each, geographers have studied the role of different groups in shaping their landscape into characteristic forms or cultural regions, marked by landscapes typical of the group in question. This in turn will raise questions about the relationship of 'culture' to people. The chapter will then look at how we might interpret such landscapes through the idea of a palimpsest. This will bring together the development of landscapes through time and the spatial diffusion of culture – the spread of ideas, practices and techniques. Such issues form the last section concentrating on the movement of cultures between the New and Old Worlds.

The changing face of the earth

The first issue to address is what is meant by landscape and what role it has played in cultural geography. Landscape above all implies a collective shaping of the earth over time. Landscapes are not individual property; they reflect a society's – a culture's – beliefs, practices and technologies. Landscapes reflect the coming together of all these elements just as cultures do, since cultures are also not individual property and can only exist socially. Much research has looked at how the landscape shapes and is shaped by that particular social organisation. This draws on an ancient geographical tradition known as chorography: the study of how landscapes bring different processes together in unique patterns. It is thus often suggested to be an *idiographic* approach – in that it is less interested in general laws than individual and unique outcomes of combinations of circumstances. The founding figure of the Berkeley school approach, Carl Sauer, considered this in a 1925 essay entitled 'The Morphology of Landscape'. He suggested geography had to start not from some idea of spatial laws, derived in some manner from the natural sciences, but from the basic experience of areal differentiation. Geography was thus based around the diversity of landscapes as 'naively given sections of reality – not a sophisticated thesis' (Sauer 1962: 317).

Sauer was not arguing for empiricism, that is, merely collecting facts about places. Instead he was arguing for a science that asked how individual landscapes came to take on their shapes. The analysis would be rigorous but there never would be some general law explaining all the outcomes. In particular what Sauer was criticising was a very influential school in the early twentieth century, led famously in the United States by Ellen Semple, a school based around *environmental determinism*. This school saw the development of cultures as a process of human adaptation to basically climatic factors. This approach has come under relentless criticism since the 1920s on many grounds – not least its incipient racism. In essence it sought to explain the different cultures of the globe through a neo-Darwinian response to environmental stimuli. Thus it suggested the northern hemisphere temperate regions had 'naturally' achieved the greatest cultural and economic development because the climate forced the populace to work, but rewarded such labour – unlike the tropics, where it suggested people had not needed to labour, and the extreme north where existence was so marginal the possibilities for accumulating wealth were limited. As such it formed a self-serving justification for European imperialism, making a process of political conquest appear to be a natural order (see also Chapter 6). Sauer was especially antagonistic to this theory. To him it went against evidence about the diversity of cultures and subjected them all to a monocausal explanation:

> Geography under the banner of environmental [determinism] represents a
> dogma, the assertion of a faith that brings rest to a spirit vexed by the
> riddle of the universe. It was a new evangel for the age of reason which
> set up its particular form of adequate order and even ultimate purpose.
>
> (Sauer 1925, in 1962: 348)

What prompted his scepticism about many such theories was not just an
understanding of the complexity of many cultures, but also a dislike for
approaches that reduced this complexity to only one factor driving the
whole system. Sauer retained a scepticism of any theories that looked not
at the region as a whole but saw it as a system producing certain isolable
end products. To Sauer the region, as expressed in its landscape as an
ensemble, was the end product. Thus either monocausal explanation or
breaking the landscape into particular products to look for 'scientific
laws' seemed misguided since 'the complex reality of areal association
was sacrificed in either case to a rigorous dogma of materialist
cosmology' (Sauer 1962: 321).

Sauer thus appealed to an idiographic idea of geography – that is, study
the unique configurations of land and life rather than seeking general
laws, the so-called *nomothetic* approach (see also Chapter 7). He
suggested focusing on the landscape as a synthetic vision, capturing the
whole operation of a local culture. Sauer felt nomothetic approaches lost
this sense of the living totality of culture by breaking it into factors and
elements. Thus in a remarkable episode, Sauer was sharply critical of the
Rockefeller Foundation's promotion and funding of high-yield maize
varieties in Mexico in the 1940s. He was so critical as to be called 'an
antiquarian' seeking to preserve old varieties as curios or museum pieces,
and locking the local populace in the past. Sauer for his part was more
obviously concerned that the local varieties of maize formed part of a
highly developed local system, and he was thus wary that it could not be
treated as one variable or tampered with without causing profound
changes elsewhere. Moreover, his studies of the origins of domestic
plants had led him to consider the diversity of varieties as a defining
feature of 'cultural hearths' – the centres of innovation. He feared that
imposing one Western grain would destroy this diversity bred over
centuries and able to cope with specific ecological and cultural niches.
He was thus talking of biodiversity before the idea became fashionable.

It may seem odd for a cultural geographer to be arguing over crop
species and gene pools, but for Sauer these were very much a part of
culture. They represented the material expressions and embodiments of
social processes and knowledge. His typically pithy summary of the issue

is worth noting, when he commented that 'if pack trails are geographic phenomena, the pack trains that use them are also' (1962: 369). That is, the materials used in, and the knowledge and skills that enable, say, planting and harvesting are just as much part of a culture as knowledge and skills about, say, writing or the structure of social beliefs. Indeed they are often profoundly linked – hence Sauer's dislike of looking at isolated factors. For instance, take the first known examples of writing in Mesopotamia: the clay tablets appear to be a tax or tribute record about a grain harvest. The farming practices of ancient Mesopotamia and the rise of dense permanent agricultural settlements have to be seen in the light of technologies of writing, and the control and storage of knowledge by elite elements, enabling the extraction of a surplus to feed the first urban settlements. What this suggests is that the issues of skills and knowledge must be seen as part of a whole system that shapes a particular landscape (Box 2.1).

Notice how this definition binds together the material and the symbolic. So, as we saw above, the knowledge and skills of one generation may be embodied in the strains of crops it produces and passes on to the next – they are artefacts of cultures. Likewise landscapes are seen as both a product of cultures and as reproducing them through time. Sauer's work suggests that artefacts may be agents of change alongside the people that use them – the tools are not simply products of people but also help shape what those people do. It is perhaps obvious then why Sauer distrusted approaches that saw 'independent' factors and variables and also why he turned to the idea of landscape and cultural regions to get round this.

Box 2.1

Cultures, their material and reproduction

A useful summary of this position was provided by the anthropologist Alfred Kroeber: 'Culture consists of patterns, explicit and implicit, of and for behaviour acquired and transmitted by symbols, constituting the distinctive achievement of human groups, including their embodiment in artifacts; the essential core of culture consists of traditional (i.e. historically derived and selected) ideas and especially their attached values; culture systems may on the one hand, be considered as products of action, on the other as conditioning elements of further action' (Kroeber and Kluckholm 1952, cited in Zelinsky 1973).

Regional personalities, cultural areas and landscape

For Sauer, the cultural region and its matching landscape was thus to form the cornerstone of analysis, forming the 'unit concept of geography' and defined as 'an area made up of a distinct association of forms, both physical and cultural' (1962: 321). It is a level at which the interaction of all the parts can be seen as a whole, but is equally defined against other areas where a different landscape can be found. So 'the unit of observation must therefore be defined as the area over which a functionally coherent way of life dominates.' (1962: 364). This sense of an integrated area corresponds with the work of Vidal de la Blache and the *Annales* school in France, where they sought to identify a *regional personality* or a *genre de vie*, expressed in the landscape. Sauer wrote approvingly of their regional monographs espousing 'the cultural landscape as the culminating expression of the organic area' (1962: 321). Again there is an emphasis on looking for the different cultures around the globe, and looking at their individual forms as a synthetic whole. The region was thus defined, not from physical characteristics as was typical in pre-war British geography, but from the way of life organised across those features (see Figure 2.1). The cultural region would almost inevitably not map neatly on to the physical, since most cultures centred on the borders of different physical ecosystems so they could utilise both (Sauer 1962: 364). In this Sauer drew on some old approaches to geography going back to Von Humboldt and beyond. Thus at the outset of his 1941 essay, 'The Personality of Mexico', Sauer states:

> This is an excursion into the oldest tradition of geography. For whatever the problems of the day that may claim the attention of the specialist and which result in more precise systems of inspection and more formal systems of comparison, there remains a form of geographic curiosity that is never contained by systems. It is the art of seeing how land and life have come to differ from one part of the earth to another.
>
> (Sauer 1962: 105)

The conception of personality here is that of a particular social system embracing the whole dynamic of land and life. In this sense Sauer is not suggesting a totally personal art, and thus demurring from the Italian philosopher Bernadetto Croce's view that 'the geographer who is describing a landscape has the same task as a landscape painter'. Rather than seeking to capture a particular view on landscape, Sauer advocated seeking out the typical or generic landscape that went with a particular culture.

Figure 2.1 A typical farm layout preserved in Ande's Zorn's Gammelgaard, from Dalarna, Sweden. The pattern of buildings is typical to the area and integrated into a seasonal rhythm with summer pastures. These buildings were built over several hundred years, and comprise typical forms and functions for a rural household with an outside cookhouse at far right and later watchmakers workshop at far left. In the middle is the typical well and store houses. The buildings go on to form an enclosed courtyard entered through arches.

However, what makes up the distinct personality of a region is not just an assemblage of parts but the way they are put together. The geographer thus would delve into particulars before returning to the synthetic level of the region. Thus Sauer's account of pre-Conquest Mexico suggests two typical cultural landscapes; one central/southern cultural region contrasted with a northern one. The dense pattern of villages, or *pueblos*, with highly intensive land utilisation in the central region supported great cities (often larger than their European counterparts) through a system of trade and surplus extraction. In contrast, there were already abandoned and ruined cities in the north by the time of the Spanish attack, where 'barbarians' had encroached and established a very different cultural system incapable of organising non-agricultural urban centres.

This typification of Mexico at the time of the Conquest suggests the elements one would look for in a cultural system. There would be distinctive arrangements of people and land, of basic livelihood, central assumptions about what is valuable or proper and thus aspirations – if not actual performance. Thus high Mexican culture was marked by intensive maize production, which sustained urban dwellers and those with non-agricultural livelihoods, which in turn required the acceptance of an Aztec elite and their extraction of surplus. This pattern enabled the Spanish to set up a colonial system of surplus extraction on the back of the Aztec system in central Mexico.

Others have looked at the invasion and colonisation of north America by Europeans for further examples of cultural regions. Thus Meinig's (1986) account of the colonisation of the Eastern Seaboard looked at the different cultural areas there. Thus French Acadian settlers had a particular sort of regional personality corresponding to a peasant culture, with subsistence production, land reclamation and dispersed settlements that was markedly different from the trading and entrepôt areas set up as part of the fur trade. Equally Zelinsky (1973) has charted these different cultural areas through the persistence of various distinctive traits – such as house or barn styles. Likewise there have been long-standing arguments over the development of a frontier culture, with suggestions that this led to a flattened social hierarchy and a culture of personal achievement linked to various strains of Protestantism. These cases illustrate the process of both adapting a culture to a new land and shaping that landscape through various cultural preferences. The starkest contrast is perhaps this individual settler model of dispersed farmsteads with the plantation landscape of the tobacco and cotton lands to the south, with the obscenely hierarchical relationships involved in slavery. At this point, though, we need to think of the two parts of these interpretations where, on the one hand, they seek to develop ideas of cultural regions and landscapes, while, on the other, they chart the spread and change of cultures. Let us consider these parts in turn.

ltural areas as a 'super-organic' metaphor?

A controversy over the Berkeley school approach to cultural areas developed from the late 1970s. Principally, Sauer is charged with treating culture as a 'super-organic' actor. That is, culture was treated not just holistically but as a single entity, as the region became too easily equated with a single actor without internal differentiation. To illustrate the problems with this we might ask whether it is justifiable to view the cultures of oppressed groups, be they the enslaved blacks in the USA or the colonised Indians, as being part of the same culture as their oppressors. This is important when we consider how black slaves were forcibly dispossessed of their own names (given their owners' surnames and christian names chosen for them) and how they struggled to develop their own culture in spirituals and other rituals. Can we say Amerindians who were having their religions destroyed as pagan or devilish products were part of one culture with the missionaries? The idea of a morphology as an 'organic or quasi-organic quality' of completeness (Sauer 1962: 326) tends to obscure these power relations.

This becomes especially problematic in work on contemporary or urban societies. The first question is about sub-cultures and their relationship to each other and a larger whole. Thus Zelinsky (1973) argues that all the sub-cultures around different artefacts, different cultural meanings and forms in the United States can still be grouped into a whole around the central values of individualism, market economics and so forth. He cautions against applying statements about the whole to any particular individual – which would be an example of an *ecological fallacy*, assuming what applies to the group as a whole applies to each member. But the central problem is that culture is both of individuals and beyond them. Sauer did not advocate an uncritical use of the 'organic analogy' but only a working device that seemed to help in the cases he studied: '[m]orphologic study does not necessarily affirm an organism in the biologic sense . . . but only organized unit concepts that are related' (Sauer 1962: 326). The question is whether the biological metaphor helps or whether it does not obscure the power relations within and between cultures. Moreover, culture is not always organically created, but can be invented or promoted or imposed as will be seen in Chapter 3 which will expand on how some different approaches see culture. The landscape or regional model also tends to downplay individual human agency – by focusing on the collective shaping of the landscape. The unit of analysis is the area or region or landscape not the actual living human beings. Equally it does not gel very well with the rapid changes of culture in urban societies. There are, though, connections with the school of urban ecology, analysing urban sub-cultures as found in discrete territorial spaces in the city. However, these studies do force us to think how the landscape can record change over time as cultures evolve and leave their own distinctive traces which accumulate into a palimpsest.

Landscape as a palimpsest

The term palimpsest derives from medieval writing blocks. It refers to where an original inscription would be erased and another written over it, again and again. The earlier inscriptions were never fully erased so over time the result was a composite – a palimpsest representing the sum of all the erasures and over-writings. Thus we might see an analogy with a culture inscribing itself on an area to suggest the landscape as the sum of erasures, accretions, anomalies and redundancies over time. As Sauer (1962: 333) put it, 'We cannot form an idea of landscape except in terms of its time relations as well as its space relations. It is in continuous process of development or of dissolution and replacement'.

There are obvious echoes in the local history approach of Hoskins (e.g. 1955) or the historical geography of Darby (1948) in the UK. In each case the landscape is the record of change, as cultural values change so new forms are required. Thus we can look at the feudal peasant system inscribed on the landscape of the open field system, marking the ox-drawn plough skills in the ridges and furrows, the relationship to the land in the collective management of the fields and the nucleated settlements. Equally we can see the rise of yeoman farming and mercantile interests in the enclosure of these fields, the spread of hedgerows and sheep leaving ridges and furrows as fossils in the landscape. The pre-existence of enclosed field systems in the south-west of the UK tells us the social structure there never totally fitted the Midlands feudal three-field model. The pattern of accretion, change and redundant forms suggests a lot about the evolution of the landscape and the local culture. Again it implies a landscape shaped and shaping the people living there, becoming a bank of cultural memories – some still in use, others as residues of past practices and knowledges. Above all it emphasises the link of people and land. A great deal of argument has gone on over how to look at such a palimpsest – as a series of layers or as temporal process. It remains a useful starting point in conceiving of a landscape, but again it tends to regional types rather than individual actors. Moreover as a temporal account of a place over time, it has to be set in terms of the second strand of landscape interpretation – the spatial diffusion of change.

ultural diffusion

Geographers have been fascinated by diffusion. Hagerstrand and others studied 'innovation' spreading among a static population, and tracked particular innovations at the scale of individual adopters. The Berkeley school was perhaps more interested in the movement and adaptation of cultures alongside specific artefacts and focused on general change rather than individuals. What this has given us is a series of rich accounts, especially focused around the European invasion and conquest in the Americas. This was a prime example of innovation and the reshaping of the landscape, origins and transformations, and evolution resulting cultures in historical and geographical setting.

Zelinsky (1973) charted the complex pattern of different types of settlers, with different cultural 'freight', arriving in different parts of the Eastern seaboard. The Acadian settlements mentioned before represent a good example of the sorts of analyses of origins that can be undertaken. The Acadians settled in an area where the indigenous peoples were non-agricultural so there was less of a conflict over land; what is more they reclaimed land from the salt marshes rather than clearing forests. The landscape created by them around the Bay of Fundy reflected a particular peasant society that drew on the agricultural knowledge and practices the settlers brought from Poitou and Aunis in France where they had seen the reclamation of marshland on the French Biscayan coast – but this in itself was a technique imported to France from the Netherlands. The movement of techniques was thus inscribed on the landscapes created. Not only that but where settlement areas were often refashioned on the model of the home region, trading landscapes were notable for being unlike the home country's – perhaps explained in the interest of trade in seeking out what is unobtainable at home.

This then exposes not only a series of different landscapes of settlers and traders in different areas but also a series of different contact zones with the indigenous population. For instance the landscape of the plains, of the buffalo-hunting tribes, was changed ahead of European invasion, by the diffusion of the horse and more efficient weapons from the south. Likewise, the First Nations of Canada were drawn into a trading circuit based on beaver pelts before they were colonised. This trade became so lucrative it sparked off territorial clashes among peoples and led to tribes attempting to dispossess others of their land in order to secure the beaver population – intensifying local conflicts, transforming the objectives and stakes in such disputes, and, as firearms spread westwards, increasing the means for violence. The gradual hunting to extinction of beaver – in river basin after river basin – to feed the European market led the hunters, and thus the conflict, to spread ever further west.

Importantly the 'actors' in all this do not appear as 'super-organic cultures'. In describing these contact zones as heterogeneous and hybrid forms, this is one of the first approaches in geography to look at the change in cultures as groups interact. Furthermore, it allows a role for non-human agents. Thus the 'firearms' that percolated westward and the horses arriving on the plains were all agents of cultural change. What are normally treated as objects or products of culture, its artefacts, are shown to be very important agents of change. More unusually it also allows a role for micro-organisms. No one can look at the history of European invasion of the Americas without recognising the significance of an often prior invasion of European diseases that depopulated areas, destabilised cultures reducing their capacity to resist invasion and often challenged the religious and local authorities – a position of authority that European missionaries then could step into. In all these elements, then, this diffusional approach seems to give a sophisticated view in many ways of how cultures might be seen to spread and change. We can see this in the example of the plantation landscape in the Americas.

Plantations, people and products

The plantation landscape represents the coming together of a web of technologies and cultures to form a characteristic pattern based on highly unequal land control, matched with an orientation to export crops, embedded in a global system of extraction, and sustained by a impoverished and often enslaved workforce. But this landscape did not appear out of nowhere or all at once. Initial European experiments in tropical agriculture were fragile indeed and often failed. It was the Portuguese who pioneered the plantation – not in the Americas but in the Atlantic islands off the African coast. It was they who, in the face of huge settler mortality rates, began to use African slaves. The Portuguese presence in the islands, especially Cape Verde, with local and mulatto (mixed ethnicity) populations gave them access to the intra-African slave trade, which they transformed in scale. By 1600, estimates suggest that these islands had extracted over 275,000 slaves, with many for the plantations on the islands but at least half sent on to the Americas and perhaps 50,000 sent to Europe (Meinig 1986: 24). It was on these islands that enslaved black labour, sugar cane and the plantation system were brought together. The islands then provided the model and testing ground of the plantation landscape as it emerged in the Americas. It was a very different model from the one the English state was developing through its 'plantations' in Ireland, sending settlers loyal to the state in order to exert control, reallocating lands to create Anglicised administrative districts, and carve up the regions of Ireland in new ways. Such plantations proved difficult to transpose into the Americas – as shown by the rather feeble British attempts to do so – though one might argue this plantation landscape finds its echoes two hundred years later as Jefferson mapped North America into geometric parcels of land for an agrarian settler landscape (see Chapter 7).

nmary

This chapter has attempted to illustrate some of the original approaches to the landscape. It has emphasised the concern with a holistic approach, and possible pitfalls stemming from that, and the connection to material culture. Also, both the *Annales* and Berkeley schools were concerned with the *longue durée*, that is changes over a lengthy periods of time – making both difficult to apply in sites of rapid change. The later sections on the diffusion emphasise the mixing and changing of cultures and the changing patterns they impose on the landscape. This does begin to challenge some ideas of organic people–land relationships, though. For a start, it suggests the need for ideas about social power, about the role of the state, and about the rapidity of change and the circuits and connections of different groups. For instance, it is possible to see an Atlantic circuit of trade sustaining an Atlantic working class, to see the commonalities and communication of ideas around the trading routes of the Atlantic as creating a disparate but coherent culture. In Chapter 10, different approaches that look to deal with these issues are outlined: rethinking culture in an era of global communication, of rapid and often perennial human mobility, of vast metropolitan societies mixing people from many origins. Here cultures may not so much be regional personalities evolving over time, but rapidly changing sets of relationships. Relationships may not form bounded organic spatial areas but may be between distant peoples or by multiple cultures existing in the same place. In order to deal with ideas of power, the next chapter looks at how the landscape can be deliberately shaped and represented to create meanings and symbols.

Further reading

Duncan, J., (1981) 'The Superorganic in American Cultural Geography', *Annals Assoc. Amer. Geogr.* 70: 181–92.

Hoskins, W. (1955) *The Making of the English Landscape.* Penguin, London.

Ladurie, E. le Roy (1974) *The Peasants of Languedoc.* University of Illinois Press, Urbana.

—— (1981) *The Mind and Method of the Historian.* Harvester, Brighton.

Meinig, D. (1979) *The Intepretation of Ordinary Landscapes.* Yale University Press, New Haven.

—— (1986) *The Shaping of America: A Geographical Perspective on 500 Years of History.* Yale University Press, New Haven.

Sauer, C. (1962) *Land and Life: A Selection from the Writings of Carl Sauer,* ed. John Leighley. University of California Press, Berkeley.

Thomas, W. (ed.) (1956) *Man's Role in Changing the Face of the Earth.* Princeton University Press, Princeton, NJ.

Zelinsky, W. (1973) *The Cultural Geography of the United States.* University of California Press, Berkeley.

The symbolic landscape

- Geopolitics: writing power on the landscape
- Relationships of inclusion and exclusion
- Iconography and symbolism in the landscape

In the last chapter we saw the landscape interpreted as shaped by the energies and practices of peoples to accord with their culture. This chapter takes a closer look at the landscape as symbolic system. That is, how it is shaped according to the beliefs of the inhabitants and the meanings invested in that landscape. In this chapter, then, we shall look upon it as a signifying system showing the values through which a society is organised. In this sense landscapes may be read as texts illustrating the beliefs of the people. The shaping of the landscape is seen as expressing social ideologies, that are then perpetuated and supported through the landscape. This chapter will start from the most intimate of spaces – that of the household – and examine how the form of the household and its relationship to the world can be seen relating to beliefs about social life. Such a view of the household will link the *cosmology* of people with the material shaping of their landscape. In the second section the quintessentially English landscape of country houses and parks will be examined in terms of the contested and changing meanings that underlay the relationship of house and grounds. From this the next section will examine how royal palace landscapes in medieval China united cosmological beliefs with geopolitical imperatives for the rulers. The fourth section will suggest how this continues in deliberately created symbolic landscapes – looking especially at the reshaping of places to convey ideas of nationalism.

House form

It is very easy to think of homes as 'natural'. They are something with which the inhabitants become so familiar they become taken for granted. However, just because something is an everyday landscape does not mean it has no meaning. On the contrary we might look at it as being the outcome of a whole set of routine practices that give meaning to daily life. To illustrate this we can look at different forms over time or over space.

Western homes and social divisions

Over time, we can see how the sort of practices involved in 'homes' have changed. If we look at the West we might characterise the last three centuries as being about a process of segregation and division. For instance, the medieval merchant's house, with a front/commercial room abutting the street, with storerooms above or behind, then 'family' rooms above that and above that perhaps workshops, was an integrated space of industry and domestic life. In differing places and times, commercial work moved to factories. Different forms of work moved at different times – affecting the relationship between genders and the values accorded to their work. The outcome profoundly structures contemporary Western life, where 'productive' labour, i.e. that which counts as 'economic', occurs outside the home while the 'reproduction of labour', feeding, clothing yourself, sleeping or childcare occurs in domestic settings. Such a division is a situated historical and geographical arrangement embodying a cultural geography where activities in different spaces are accorded different status and economic values. So the home may be seen as part of a gendered landscape, one that has served to maintain the idea of a working man's wages, as the 'breadwinner', and maintain the idea of a 'woman's realm' in the home. Such landscapes have been shaped and reshaped of course and it does not do to be too sweeping about all this. Thus if we look at the British town house, we can see great changes in the 1930s and into the post-war era. The size of dwelling declines and its internal shape changes as economic and cultural changes occur about what comprises a family unit. It is vital to remember that until the First World War the badge, almost the definition, of being middle class was to employ a servant. So town houses were arranged with this in mind, with servant rooms in the attic or 'downstairs' and out of sight of guests. The maintenance of the household, the food preparation, the laundry and so forth, was hidden away in these quarters. With the decline of domestic service, the modern house becomes designed more for the efficiency of such tasks rather than to hide them away.

The routine spaces of homes speak to us about the sort of social relationships we believe in and the practices that sustain them. We can think how much practices of separation have come to constitute the Western idea of a proper home. Economic activities occur elsewhere. The decline of servants means the house is often inhabited by a family, a kinship group, on their own. Within its very structure the spaces of display to visitors, 'front rooms' and best furnishings, are separated from those of everyday life and rest – the bedrooms (see Figure 3.1). Indeed we can chart through the last two centuries the changing moral geographies in first the separation of sleeping quarters from living, then the separation of adults from children and the separation of children by gender. Judgements about morality and sexuality are written into the fabric of the house through the creation of private spaces.

‪byle housing

We can put Western arrangements in context if we look around the world at other peoples. We could look to Malaysia where the Dayak's of Sarawak traditionally lived in longhouses that contained far more than the individual family group. As a detailed example we shall look at the Kabyle of Algeria based on work by Pierre Bourdieu (1990). Their dwellings tend to contain an extended family group in a covered single-storey rectangular building, along with spaces for weaving, storing agricultural produce and fodder, and indeed stables for animals. The arrangement of these activities can tell us about the world-view of the Kabyle themselves, how their cosmology structures their daily practices (Figure 3.2). The house tends to be on a slight incline to allow drainage; a slope that structures activities so the down-hill end contains all that is damp, dark and green – also then becoming the place for natural human activities of birth, sex, sleeping and death – while the upper end contains all the activities associated with light, fire and entertaining guests forming a division of the civilised and natural. A slighted guest will thus complain of being made to sit against the dark wall of the house. Women's work tends to occur in the dark sections of the house while male labour occurs outside. Thus the house with its sets of oppositions of man and woman, light and dark, high and low, natural and civilised also

Figure 3.1 'L'esprit en la virilité' *(The Spirit of Virility) by Abraham Bosse, c. 1630. In this picture of a well-to-do family in seventeenth-century Paris there is nothing surprising in having a meal in the same room as beds. The separation of these activities in the West spread socially and spatially so that the rural peasant houses in Figure 2.1 still have combined sleeping, dining and cooking areas in the nineteenth century.*
Source: Bibliotheque Nationale, Oa 44, pet.fol., p. 22.

interacts with wider cosmologies. Thus men will leave the house before dawn, so the outside is male space and the inside is feminised. Male friendships are then described as 'outdoor friends'. The house is then thought of as separated against all the outside world. In short the oppositions that structure its internal spaces structure how it relates to the outside world:

> Considered in relation to the male world of public life and farming work, the house, the universe of women, is h'aram, that is to say, both sacred and illicit for any man who is not part of it.
>
> (Bourdieu 1990: 275)

Bourdieu observes that the house itself is divided according to the principles that divide it from the outside; the same oppositions organise both. It is thus possible to look at the spatial arrangement of the landscape and the practices that shape it in order to look at the cosmologies of both ourselves and others. There is no simple natural disposition of activities in the landscape, they are always bound up particular cultures. We have then a geography on two levels, the way cultures use geography, investing meanings in certain spaces, and then (but not only) the geographical distribution of these cultures.

house and the garden: English country houses

This section looks at how a familiar landscape uses space to embed certain meanings by looking to English country houses, and thus links to material discussed in *Historical Geography* in this series. The English country house has been used to symbolise the very heart of English national identity; indeed enthusiastic commentators have gone so far as to suggest it is *the* contribution by the English to global civilisation. It has been used as a talisman for a conservative vision of organic rural values: a landscape of squires and reciprocal relations between classes that has been consistently mobilised as the opposite of state welfare, contrasting the personal attachment of people and places, the way people are known and know their place in these landscapes with the impersonal bureaucratic welfare state. If such landscapes are at the 'heart' of England then their spatial arrangement says a great deal about the values that have formed that heart, and the political connotations of that landscape. These are not neutral expressions of innate values; they are social landscapes that tell us about the social relationships and beliefs in society.

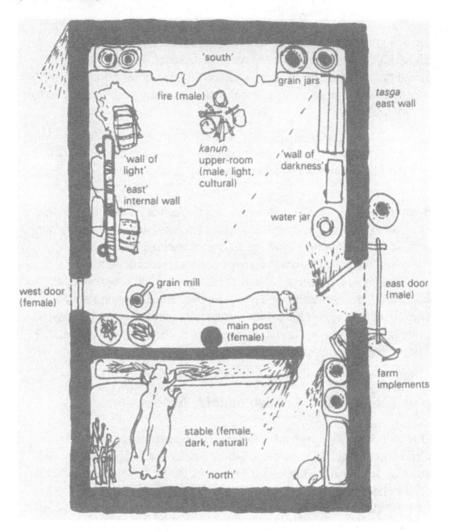

Figure 3.2 *Plan of a Kabyle house*
Source: adapted from Bourdieu 1990 by Oliver 1997.

Perfectible nature

There is a substantial literature on garden history and its relationship to prevailing social understandings. Here we can only tease out a few suggestions and examples. From the medieval period the garden was designed as a place of contemplation and 'earthly delight', but how this has been expressed has changed over time. For instance, in the William and Mary period in the late seventeenth century, garden design was marked by geometrical patterning. Now the idea of using geometry on large gardens produced layouts of radial avenues symbolising power emanating from the house at the centre. In more modest gardens and those near the house it was expressed in an extremely formal layout, with flower-beds in geometric, often rectilinear shapes, sharply marked off by paths. Hedges were often planted, as small box hedges to separate different flower-beds, or as larger creations and, along with trees, as objects for topiary – trimming into cones and angular shapes was common. What does this say about how people thought about nature? The geometric regularity and order of the garden seem to express a stark contrast with ideas of a wild nature, often signified by walling gardens off from the rest of the world. Walls and bounding hedges thus play a significant role:

> The highly structured and artificial patterns in the gardens they surrounded made the best visual sense when clearly segregated from the less ordered environments around them. To the modern observer these gardens appear as places where nature has been ordered, tamed, even tortured – quintessentially 'unnatural'. Whether contemporaries would have seen them quite like this is less clear. Their more educated owners would perhaps have explained them in Neoplatonic terms, as expressions of the perfect forms underlying the imperfect shapes in the natural world.
>
> (Williamson 1995: 31)

Neoplatonism is a view of nature where humanity is seen as having a duty to reveal the divine order behind nature. Geometric layouts were thus not opposing nature but striving to perfect it, or to bring out the perfect essence within it.

rkland and prospects

As beliefs altered so too did garden form. Thus, as the eighteenth century began, an increasing proportion of grand gardens was set aside as 'wilderness' expressing a new relationship to the land; visual command of property through the idea of the vista (that is, the appearance of an object at a distance). Such vistas were created by opening up alleys through woods to reveal the spires of churches or buildings in the distance. The role of this can be seen if we consider Moseley Wood where Cockridge Hall has walks laid out through the woods that provide 65 intersections and 306 different views. One significant feature was the rise of the ha-ha (a sunken ditch at the edge of a garden or lawn). When cut into a hill, the garden is on the upward side and separated from the outside world by a wall or bank below foot level that leads to a ditch before the outside land. As such it is designed to stop animals wandering into the garden. In contrast to seventeenth-century walled gardens the ha-ha is invisible and nothing interrupts the sweep of the eye from garden to outside land – which is included in the overall view rather than separated.

The significance of this rather obscure garden feature is, then, the visual mastery offered. The owner of the garden no longer sees their 'patch' set off from an uncontrolled outside world; rather this was an expansive owning vision – conflating visual and social mastery. We can connect the rise of this visual mastery with the creation of 'natural settings' for country houses – the 'park land'. The removal of perimeter walls illustrates the continuing rise of the park and the growing importance of a naturalistic setting for the mansion (Williamson 1995: 47). This landscape was shaped by the struggle to control physical and visual access. There are numerous cases of villages, cottages or farms being moved to make the gentry sole occupiers and sole possessors of the scene. So although the removal of walls around properties might seem to mark an opening out of a landscape, country houses were still set in landscapes shaped by exclusions.

Polite society, power and exclusion

These parks corresponded with the mania for hunting among the gentry, forming game larders as well as grazing land. The rise of woodland in part reflected the rise of pheasant shooting – with many small woods planted as estates competed in terms of the total slaughter they could create. This competition continued into the nineteenth century and meant that exclusive shooting rights, the protection of game from unauthorised people, became more important. At the same time the creation of exclusive parks and the consequent immiserisation of the peasantry created considerable conflict. The traditional rights of rural folk to gain sustenance off common and wild lands were being removed, and replaced with the exclusive hunting rights of the gentry. A measure of the conflict's bitterness is how the Poaching Game Act (1770) meant, on the word of one witness, anyone going about the woods at night could get six months gaol; that of 1773 meant that a second offence could lead to a public whipping; and by 1800 gamekeepers could arrest people without a warrant if in a group of two or more, the perpetrators would then be classed as 'incorrigible rogues' and subject to two years in gaol, whippings or being press-ganged into the armed forces. Perhaps most telling of the scale of struggle over this landscape is that one-sixth of all the convictions in England at the beginning of the nineteenth century were for game offences.

Exclusion and conflict were thus symptomatic of the landscape of polite society: 'The mansion thus lay in the midst of an insulating sea of turf, hidden from view by encircling belts [of trees]. And once established as a sign and symbol of exclusivity, the patterns of social contact which the park engendered could only serve to perpetuate the emerging divisions in rural society' (Williamson 1995: 102). The turnpike roads formed the arteries of this polite society, as gentry moved from park to park observing the intervening country from within a coach. Such a practice symbolises the divisions in rural life on which the country house was founded. Embedded in these avenues is a politics over access. To ensure seclusion in the park, roads would be closed – a process that, after 1773, involved just two magistrates, who were generally of the same social set as the landowner anyway. The divisions these landscapes were founded upon can be found in the writing of a poet, from Bedale in Yorkshire, about the local Rand manor house:

> And now them roads are done away,
> And one made in their room,
> Quite to the east, of wide display,
> Where you may go and come,
> Quite unobserved from the Rand,
> The trees do them seclude,
> If modern times, do call such grand
> It's from a gloomy mood.
> (Hird, cited in Williamson 1995: 106)

Thus Humphrey Repton, the landscape gardener, argued for the use of woods around the edge of the property – both to seclude and, on smaller parks, give the impression of added depth and distance belonging to the park. These trees formed objects of beauty and profit, symbols of ownership and nationalism. For a start, the return on trees was low, though on marginal land it was as good as grazing, so it symbolised having the cash to take a long-run view. Likewise Britain was terrified of running out of oak, especially for the naval dockyards, so planting oak was a patriotic investment in the future of the nation. To succeed it also required exclusive property rights and allowed the rearing of game. The park and the trees form part of complex constellation of meanings and values.

Moreover in the changes from the landscape of geometric patterns we can see the evolution of society. Social relationships were becoming more fluid *within* the gentry, even as they excluded the rural poor from the scene. Social events had been very hierarchical, involving the presentation of various people to the hosts according to status. By the eighteenth century, such formality was declining; people would expect to 'circulate' between activities – cards, dancing or conversation. The park thus became a landscape allowing slowly changing views and the circulation of people rather than the fixed and ordered viewing points described earlier. The social vision of a cohesive polite society was part and parcel of a vision that wanted exclusive ownership.

Geopolitics: writing power on the land

The sacred landscape

A different example of landscapes purposefully shaped to reflect cosmological visions and related to geopolitical situations can be found in the ancient Chinese summer palace of Chengde. This was built between 1703 and 1792 by the Manchu emperors who succeeded the Ming dynasty. The very location of Chengde is north of the Ming heartland and north of Beijing, reflecting the new powerbase of the Quing empire, which centered on Manchuria and Jehol, and the expansion of that empire to both sides of the Great Wall. The landscape itself was shaped by explicit beliefs in geomancy – the magic powers of the earth and in *feng shui*. Thus the 'male' mountains surrounding the site are balanced by creating 'female' elements of gardens and lakes. These lakes formed eight pools and nine islands, echoing the Buddhist proverb that the world consists of nine mountains and eight seas. The idea of a universe consisting of concentric mountain ranges leading to a central mountain, Mount Sumeru inhabited by Indra, is reflected in the erection of a central, artificial peak crowned by a temple. Forêt (1995) argues the palace showed a non-Chinese dynasty trying to establish a geopolitical claim over the diverse territories of an empire. Key symbolic elements of other centres were brought to the new palace from Beijing, Lhasa, or Wutai Mountain; the summer capital can be read as a composite landscape that reproduced the map of the Manchu empire where the order imposed within the garden mirrored the larger order imposed on the conquered territories.

ionalising space through monumental landscapes

A more recent example of creating places to symbolically bind together territories can be found in central Jakarta, which is striving to represent an independent Indonesian nation state. Indonesia was created out of a collection of principally Dutch colonies, comprising different religions (mainly Muslim, but also Hindu, Christian and others) and various ethnic groups. The task confronting President Sukarno on gaining independence was to weld one of the most populous and diverse parts of South East Asia into one state. Macdonald (1995) suggests the symbolic landscape was manipulated to support this project. The purpose of looking at the manufacture of these symbols is 'not to measure their authenticity against some historical yardstick but rather as a means to tease out the complexities of representing a viable geo-political basis for a collection of territories recently emerged from colonial domination' (Macdonald 1995: 272). In Jakarta, the colonial administration was clustered around 'Konigsplein', which became renamed as Medan Merdaka, to symbolise the Indonesian state rather than the intrusion of Europe. The former centre of colonial administration was rewritten as the heart of series of concentric circuits, centre of Jakarta, centre of Indonesia and part of a world of co-equal modern states. As such it rewrote what had symbolised European rule as symbolising Indonesian-ness in a subtle way, for while it asserted its newly won independence, it also reincorporated the heart of colonial power (important since the claims of the Indonesian state to rule its territory were based on inheriting that territory from the prior rulers), so the governor's palace smoothly became the presidential palace. It was by no means inevitable that a singular, Indonesian state would emerge – it could have been founded on Javan ethnicity, Communist liberation movements or Islamic law – all these forces were shaping the state and any could have tipped the balance. The final landscape expresses how it was that a nation state emerged on a particular model, that was legitimised though the landscape.

For instance, state power was set up as autonomous from the powerful Islamic forces in the region that had always contributed to Indonesian identity. Thus a national mosque was an obvious part of this national landscape; but there are subtle messages in its design. Unlike neighbouring Malaysia the national mosque is not built in Asian style but rather using the domed architecture of Arabist styles. The architecture thus identifies Islam with a pan-national identity, not a national one; it relocates Islam's claims away from controlling the national polity to a realm of international influences. This is further stressed by the fact that next door to the mosque remains the Dutch Catholic cathedral. Ostensibly a gesture of tolerance and reconciliation after independence, its presence counterbalances that of the mosque symbolically – suggesting that many outside world religions have played a part in shaping modern Indonesia. However, Christianity is a minor religion (and Catholicism a sect within that) compared to the prevalence of Islam; by suggesting their equivalence the new rulers also suggested Islam had no especial claim on the polity.

The centre of Medan Merdaka square is a tower, the Monas monument, raised to stand over the former colonial buildings. In the monument is a series of forty-eight dioramas, bound into a narrative through their spatial logic – simply walking from one to the next is enough to link them into a story leading up to the creation of Indonesia as a modern state. They form a purposeful sequence, chosen especially and in particular order to make the final outcome appear foreordained (what is called a *teleological* story). Thus the different forces that shaped Indonesia are given different significance by what role they are depicted as playing in this story. In the first, there is a picture of forced labour and plantation life showing the position allocated to Indonesian people in a brutal colonial system – that of agricultural producers for the benefit of the West. The next picture is also of the Dutch world order but is of the Protestant church, captioned as the 'role of the Protestant church in uniting the nation' – as though that effect was foreordained – and showing the need to reject the colonial legacy yet also to claim it to legitimate the territorial claims of the state. A whole panel is devoted to the United Nations building in New York, not the people just the building – symbolising the moment when the international community recognised the claims of Indonesia as a nation state. Likewise, a little distance from the square, is the separate monument to the accession of West Irian, now Irian Jaya, the last territory ceded by the Dutch. The monument of an individual rising up and breaking their chains is meant to symbolise the final bonds of colonialism lifted.

However, it is important not to give the impression that this reworking of themes to support a particular idea of the Indonesian state is entirely successful. The dioramas are captioned in Javanese, the language of the

dominant island, and English, the most common language of tourists –
any of the other peoples of Indonesia cannot read them. After Sukarno's
downfall, the all too obvious phallic statement of the tower in the square
became known as 'Sukarno's last erection'. Meanwhile the monument to
struggles for freedom at West Irian takes on a new meaning given the
struggles of the peoples of Irian Jaya and East Timor against the
Indonesian state to become separate nations based on their own identities
rather than be incorporated into Indonesia.

tionalising space through rewriting the past

It is not just new buildings that can be created to alter the symbolic
landscape. Ancient landscapes have been given different interpretations
over time indicating the way the meaning of places can become a matter
of political contest. Cambodia's ruling party in the 1970s, Pol Pot's
Khmer Rouge, found it useful to promote a particular interpretation of
the ancient and ruined palaces of Angkor Wat. They were sceptical of
urban groups and wanted to pursue an isolationist policy. They found
evidence of a Khmer culture existing before any Western contact useful
in bolstering their claims that they did not need links to the rest of the
world, and their policy of eradicating the legacy of French colonialism in
Indo-China. Moreover, they used the elaborate canal system as the basis
for creating an irrigated agricultural system – that failed to feed the
populace. Such a use of the symbolism of Angkor Wat helped legitimise
a policy that led to hundreds of thousands of deaths before the
Vietnamese invasion deposed the Khmer Rouge in 1979.

A different example can be found in Zimbabwe, where the ruins of Great
Zimbabwe caused symbolic difficulties for the white rulers of Rhodesia.
Their rule was legitimated by discourses or stories about the black
population being incapable of self-government and being less 'advanced'
in some sort of ladder of civilisation, and, in some quarters, as being just
as much recent arrivals to the region as the white rulers. Yet here were a
set of ruins from the fifteenth century, at least as impressive as anything
in Europe. White society dealt with the symbolism of these ruins through
a variety of means: from allegedly scientific studies through to popular
mythology and romantic histories. Textbooks during white rule thus
ascribed them to Arab traders, or some earlier people that had died out
(or been destroyed by the current black inhabitants) or even to the
mythical figures and lost 'white civilisations'. With majority rule this

changed – the ruins now having a symbolic centrality to the state mythology, and appearing as a recurring motif in national symbols, such as bank notes. The current regime can use the antiquity of a Zimbabwean polity to add legitimacy to their claims for the modern state. They can now retell the history of the ruins as the fall from a black-dominated golden age, and a current 'resurgence of our Zimbabwean civilisation' (cited in Kaarsholm 1989: 91). These three examples illustrate the role of landscapes in shaping identities for a people, in a place over time. The shaping of landscape can reflect and reinforce ideas of what constitutes a people, who is included, or excluded – so the polite society of country houses excludes the poor, while Indonesia struggled to invent an inclusive idea of Indonesian-ness. And such can involve 'inventing histories', in shaping ideas of how that people relate to their place and their past (see Chapter 10).

Summary

It should be apparent that we cannot see landscapes as simply material features. We can also treat them as 'texts' that can be read, and which tell both the inhabitants and us stories about the people – about their beliefs and identity. These are not immutable nor ineffable; some parts may be taken-for-granted parts of everyday life, but others may be politically contested. Landscapes are open to struggles over their meanings – be that the political use of cosmology in China or the contested histories of Zimbabwe. Reading the landscape is not a matter of finding a typical 'cultural area', as in the last chapter, but of seeing how landscapes come to mean different things to different people and how their meanings change and are contested.

The situation is complicated by what might be called double encoding of landscapes. That is where landscapes are wrapped in another representation. Thus in the case of country houses, their landscapes had meanings for visitors at the time they were built. Contemporary viewers can see them represented in paintings, book illustrations or television. Each of these might put different spins on the landscape – using them for particular purposes in a programme, say. So we have our own contemporary values on top of those in a landscape that was already saturated in meanings. The situation can then get very complex. To give a brief illustration, we can think of country houses by the eighteenth century as being committed to a managed landscape, one with order that could be envisioned as a whole. It was in the terms of the time an 'improved' landscape – one that showed it was cared for, and owned, by its order. Yet if we think of the paintings of, say, Constable we find them full of features which would have irritated local countrymen such as dead trees, broken gates, or a neglected flock of sheep which were aimed at urban tastes (Daniels 1993: 204). These paintings are now used to promote tourism and to signify a rural idyll away from the speed and bustle of modern urban life. The next chapter will start to think through how places and landscapes are re-presented in literature, and in Chapter 5 will look at the role of films and TV in more detail.

ırther reading

Barnett, A. (1990) 'Cambodia Will Never Disappear', *New Left Review* 180: 101–26.

Bender, B. (ed.) (1993) *Landscape: Politics and Perspectives*. Berg, Providence.

Cosgrove, D. (1985) 'Prospect, Perspective and the Evolution of the Landscape Idea', *Transactions of the Institute of British Geographers* 10: 45–62.

Cosgrove, D. and Daniels, S. (eds) (1988) *The Iconography of Landscape*. Cambridge University Press, Cambridge.

Daniels, S. (1993) *Fields of Vision: Landscape Imagery and National Identity in England and the US*. Polity Press, Cambridge.

Duncan, J. (1990) *The City as Text: the politics of Landscape Interpretation in the Kandyan Kingdom*. Cambridge University Press, Cambridge.

Forêt, P. (1995) 'The Manchu Landscape Enterprise: Political, Geomantic and Cosmological Readings of the Gardens of the Bishu Shanzhuang Imperial Residence at Chengde', *Ecumene* 2(3): 325–34.

Hobsbawm, E. and Ranger, T. (eds) (1989) *The Invention of Tradition*. Cambridge University Press, Cambridge.

Kaarsholm, P. (1989) 'The Past as Battlefield in Rhodesia and Zimbabwe', *Culture and History* 6: 85–106.

Lonsdale, J. (1992) 'African Pasts in African Future', *Canadian Journal of African Studies* 23: 126–46.

Macdonald, G. (1995) 'Indonesia Medan-Merdeka – National Identity and the Built Environment', *Antipode* 27 (3): 270–93.

Oliver, P. (1987) *Dwellings: the House Across the World*. University of Texas Press, Austin.

Pardailhe-Galabrun, A. (1991) *The Birth of Intimacy : Privacy and Domestic Life in Early Modern Paris*. University of Pennsylvania Press, Philadelphia.

Williamson, T. (1995) *Polite Landscapes: Garden and Society in Eighteenth-Century England.* Johns Hopkins University Press, Baltimore.

Zukin, S. (1991) *Landscapes of Power: From Detroit to Disney World.* Berkeley, University of California Press.

4 ▶ Literary landscapes
writing and geography

- ◦ Conveying a sense of place
- ◦ Urban geography and novels
- ◦ Modern experience and styles
- ◦ Texts about places or space in texts

Over the last twenty years geographers have become increasingly interested in various forms of literature as ways of investigating the meaning of landscapes. Literature is replete with poems, novels, stories and sagas that describe, strive to understand and illuminate spatial phenomena. This chapter will trace a series of these engagements. The first way is perhaps the most obvious, where literature about places has been used as a source or as data. Just like a survey, literature becomes another set of geographical data available for use. In the dark days of the quantitative revolution it was downplayed as 'subjective' – a representation whose relationship to a (statistically verifiable) reality was suspect and could not be tested. The starting point of this chapter is interest in 'subjective' experiences of place, how people come to understand places, and to thus identify a human geography filled with emotions about places – where places have meanings beyond their statistical expression. The skeletal landscapes of statistics miss out the richness of human experience of place. Thus the first section looks to prominent writers about regions, seeking to show the affective, emotional, relationship of people to spaces, developing from early work by such as Darby (1948) on Hardy's Wessex.

> As a literary form the novel is inherently geographical. The world of the novel is made up of locations and settings, arenas and boundaries, perspectives and horizons. Various places and spaces are occupied by the novel's characters, by the narrator and by audiences as they read. Any one novel may present a field of different, sometimes competing, forms of

> geographical knowledge, from a sensuous awareness of place to an
> educated idea of region and nation.
>
> (Daniels and Rycroft 1993: 460)

It is clear, though, that literature cannot be read as simply describing these regions and places – in many cases it helped to invent these places, so the section goes on to look at the processes whereby literature can create geographies. It is a simple point, but most people's knowledge of most places comes through media of various sorts, so that for most people the representation comes before the 'reality' (a point taken up in the context of travel writing and imperialism in the next chapter). Most people know about 'Wessex' through Hardy not through personal knowledge. Literature (along with other more recent media) plays a central role in shaping people's geographical imaginations. This leads into the second section which shows how different modes of writing express different relationships to space and mobility, and how spatial relationships within literature can be invested with different meanings. Not only does the work say something about a place, but its very construction says something about how society is ordered spatially.

Literature is not flawed by its subjectivity; instead that subjectivity speaks about the social meanings of places and spaces. Thus I consider different ways of writing about the city and what different forms of story, from different periods and places, tell us about the nature of urban life. Building on this I suggest different literary genres tell us about changing periods – how the rise of modernity, and indeed postmodernity, in literature corresponds to different ways of experiencing the world and organising knowledge about it. Finally, these different examples open up the relationships of geography and literature – suggesting more complex relationships than as source or as a subjective geography. Geographers deploy imaginative techniques and literature also engages with material social processes. Geography and literature are both writings about places and spaces. They are both processes of *signification*, that is, processes of making places meaningful in a social medium. I shall conclude by suggesting not only does literature involve such signification, but so also do geographical writings about places.

Writing about places

If anyone were to look around for accounts that really gave the reader a feel for a place, would they look to geography textbooks or to novels?

The answer does not need saying. Undergraduate geographers receive years of training which seem to remove the ability to write a piece of prose (let alone, say, poetry) that imaginatively engages its reader. Such is a slightly sad state of affairs and leaves geography a more arid, desiccated and poorer discipline. This is especially important if we are trying to describe what landscapes mean to people. Humanistic geography in particular seeks to put human experience of places back as the central concern of geography. Now this can involve getting people to talk about their experiences of places, about their lives and how they see the world. Humanistic geographers also quickly realised that accounts in literature provide similar insights into the experience of places. In such ways we might turn to novels to look at a sense of place in the evocation, or what might be called the word-painting of places.

Such evocative accounts allow geographers to look at the *genius loci*, the unique 'spirit' of a place. Such a central experience of geography is not the location (however exact), not the most elaborate enumeration of details – these do not approach the essence of meaning of a place, or as Heidegger put it 'geographers never study the springhead in the dale'. Such a concern with the meaning of landscape finds echoes in the words of Burns (currently used in a Scottish Tourist Board promotion); gazing out across the Highlands, he is heard to ask how the mind of man can map such landscapes as this into abstractions. In literature then, humanistic geographers found accounts that looked to the experience of place, where '[t]he truth of fiction is a truth beyond mere facts. Fictive reality may transcend or contain more truth than the physical everyday reality' (Pocock 1981: 11).

Interest initially focused around regional novelists who most clearly felt, and created, a sense of place through their writings. Thus in the writings of D.H. Lawrence, we find a densely worked account of life in the Nottingham coal field, and the experience of the working classes expressed through landscapes of class solidarity in towns and landscapes of freedom in the countryside. Thomas Hardy's descriptions of Wessex folk, their customs and dialects, offer a coherent regional identity. He can also be seen as writing an elegiac landscape, memorialising the end of a rural way of life. The trudging, dispirited pace of Tess Derbyfield's family on a forced migration speaks volumes for the process of social division and immiserisation; the new moneyed d'Urbervilles in their mansions can be read to add a vivid layer to our earlier account of country houses (Chapter 3). The landscape of *Tess of the D'Urbervilles* shows the power of money over the land, symbolising this through Alex

D'Urberville's power over Tess – illustrating a gendered connotation to country houses' power over the landscape.

We need not look solely at novels. Some of the best-known writers about places are poets. To continue the theme of rural decline, we could look at Goldsmith's poem of the deserted village. Here every falling-down post, every overgrown verge speaks to a sense of sorrow as an earlier rural world was destroyed by industrialisation. Poetic evocations of places and feelings can arouse high passion. Blake's poem 'Jerusalem' celebrates a vision of the heart of England as 'mountains green', contrasted to the corrupting 'satanic mills' of the industrial revolution. His fellow Romantic, Wordsworth, also wrote of those mountains in the Lake District – where he 'wandered lonely as a cloud', trying to conjure a sense of the sublime in nature. This Romantic vision of landscape sought the majesty of nature, the 'sublime' that exceeded the merely human. These poems are historic events in themselves. They were informed by the social context of the day and then themselves went on to inform that context. Thus Wordsworth popularised the Lake District and others came to seek the sublime experience he recounted. So his place evocations played no small role in shaping geographies of tourism, and later national parks and thence to agricultural practice. And this is not an isolated case. Beatrix Potter also popularised the Lake District as the site of her home.

> The literary meaning of the experience of place and the literary experience of that meaning of place are both part of an active process of cultural creation and destruction. They do not start or stop with an author. They do not reside in the text. They are not contained in the production and distribution of the work. They do not begin or end with the pattern and nature of the readership. They are a function of all these things and more. They are all moments in a cumulatively historical spiral of signification.
>
> (Thrift 1981: 12)

Accounts are involved in a never-ceasing spiral of signification, where their meaning may change as the context changes, where they draw on each other to form genres. For instance, the destruction of rural life is an idea that occurs repeatedly over time. Wherever you look the authentic rural landscape is shown as on the verge of disappearing – a true rurality is always suggested to have existed just a generation before – in a sort of ever receding escalator. We must be careful of assuming literature can get us directly to some *genius loci*. These works are not transparent accounts of a sense of place. They draw on other works, on wider philosophies, and on techniques of writing. To understand this we need to think through the specific relations of literary production in historical context. This allows us to interpret historically embedded 'structures of feeling' (Williams 1977) about a place in a specific period.

paces in the text

riting home and away: ordering spaces

Darby (1948) attempted to tie Hardy's Wessex to the social and physical 'regions' of the area – relating the region as comprised in literature to that in geography. Such a simple overlaying of one 'map' with another may be interesting but is fairly limited in scope. What is perhaps more interesting is seeing how certain places and spatial divisions, are established within the literary text. This comes through both in the plot, character and autobiography of authors:

> Home is like a fortress of an army which prides itself on its mobility . . . Departing from the base, feet define geography, the eyes observe and systematize it . . . As the base line in surveying is essential for the formation of a map and all points on it, so the connected points of birth, place and upbringing are – for any person and even more so a writer – factors never to be relinquished.
>
> (Alan Sillitoe, quoted in Daniels and Rycroft 1993: 461)

The creation of a sense of home – and homeland (see Chapter 5) – is a profoundly geographical construction in a text. Such a 'base' is vital to geographical knowledge about the imperial and modern worlds. One of the standard geographies in a text, exemplified in travel stories, is the creation of a home – be it lost, or returned to. The spatial story of many texts echoes the pattern of travelogues, with the hero leaving home, suffering deprivations, doing deeds and then returning vindicated. If we go back thousands of years, the epic poem of *Gilgamesh* – one of the earliest literary pieces from Middle Eastern civilisation, bears exactly this pattern. Homer's *Odyssey* conforms to it, and in a more bitterly realised way so too does Aeschylus's *Oedipus Rex*. We can think of fairy tales, the stories of knights and derring-do, the plot of hundreds of novels, including adventure stories and current travel accounts.

The structure though betrays some significant cultural geographies and also some gendered geographies. It might be fair to say that this structure 'domesticates' the home. The home is seen as a place of attachment and security yet also confinement. In order to prove themselves, the *male* heroes leave (either through folly or volition), into a space of male adventure. In the *Odyssey*, Odysseus eventually is forced away from his home and family to a long war and long voyage of return. In these deeds and his voyage he has been argued to exemplify classical ideas of humanity – struggling to carve out his own destiny. In the course of his travels he proves himself in battle and strategy, and on the way home continues to battle the world while sleeping with various women. He returns home to find Penelope, his wife, resisting her suitors, his son's patrimony in jeopardy and has to reassert his authority over the home. The journey has marked him out above all in a masculine space. It is interesting that of five 'Trojan war' epic poems this is the only one to survive; the others dealt with such as Agamemnon's return and murder by his unfaithful wife Clytemnestra. Homecoming can have more troubling meanings – suggesting the danger and fragility of male power in the home. Careful reading also suggests the importance of this spatial structure in creating an idea of home. The starting event is always the loss of a home. Struggles to return are thus organised around a lost point of origin. Countless stories go on to suggest that returns are rarely unproblematic. Indeed modern stories often suggest how things can never be as they were before. The notion of 'home' created through this structure might be called a retrospective fiction – a nostalgic looking back on what has been lost.

The shifting relationships of mobility, freedom, home and desire have been suggested as an allegory for a very masculine experience of space. If we look at the Beat poetry of Jack Kerouac in the 1950s, or the music of Woody Guthrie (see Chapter 6), there is a change with the celebration of drifting. The heroes do not seek a return to some stable home, indeed they reject such notions. However, we can still see the clear division of the male heroes, fleeing commitment on to the open road to escape a feminised home seen as confining them. In this case surely we are watching a gender ideology mapped through literature on to space – confining women to 'home-creating', associated with security and nurturing and ejecting men on to the road, to 'escape' to a freedom and to prove themselves. In both cases men and women are cast not only into spatial relationships, but those relationships help support what the experience of place of is, and what it means for a man and a woman – they are both assigned gendered desires through geography. Such suggests a close connection between spatial experience and personal identity. Social values and ideologies can thus be seen operating through spatial categories, moral and ideological geographies, in literature (as much as in the Kabyle house in Chapter 3).

These moral geographies can work in other ways than solely in terms of mobility. In Rabelais's work *Gargantua*, we can detect a social geography of taste and manners. Through the tales of Gargantua's desires, bodily gratifications and obscene behaviours we can chart a geography of (im)polite or (in)appropriate behaviour – the creation of an (ill-)mannered body, disciplined according to various spaces. Certain spaces being coded to allow different behaviour at different times: some for eating, others for sleeping, for washing or defecating. Spaces become coded according to manners and those manners symbolise a position in society. Such a geography of order is about a series of moral and cultural judgements of what should happen where. Gargantua's obscene behaviour reveals these norms by breaking them. For instance, Rabelais's account of early modern society offers several accounts of carnivals, fairs and disorderly marketplaces. In Rabelais's work these are spaces where social norms are inverted: where the 'fool' is appointed chief as the 'lord of misrule'; where low culture reigns higher; where the normal social world is turned upside down. They are marked by a sense of 'carnivalesque' (see also Chapter 8). In novels, we can chart the emergence of such 'ludic' spaces, that is, spaces in between rules where disorder is licensed, and previously forbidden behaviour tolerated.

Literary accounts can thus reveal something of how spaces are ordered and how relations to spaces can define social action. Such relationships occur not only at the level of region, or place, but can be the relations of home and away, forbidden and accepted behaviour, permitted and transgressive behaviour. The meanings of space in literature can be more subtle than simple place attachment. However, so far we have really only looked at the ways spaces relate to each other in texts, not at the specific textual forms and styles – that is the way characters, plots and narratives come together. The next section looks at stories about cities, the changing form of the text and how it relates to specific geographies.

ting the city

The city has long been the scene of many novels. However, it is possible to gain richer understandings than by simply using this as 'data', however evocative, about urban life. The city is not only a setting for action or stories; the depiction of the urban landscape also expresses beliefs about society and life. We have already seen how writings about rural scenes may mobilise wider ideas of decline and social change in how they talk of landscape or how the rural can be set up as a pastoral idyll signifying a virtuous social order (Chapter 3), and how writings can express a moral geography of social life and behaviour. So it is not a question of how accurately the city or urban life is depicted; rather it is a question of what the urban is used to signify, what the urban scene means.

In *Les Misérables*, Victor Hugo structures central episodes of the novel around Paris. The alleyways of the poor form a geography of imaginative darkness, a mysterious geography of an 'unknowable city'. The novel often takes an aerial view but this view does not allow a perfect knowledge of the city; the city remains dark, ominous and labyrinthine. The novel refers to an even more impenetrable geography, a literal and figurative underworld – a world opposed to officials and the state. The explosive revolt depicted in climactic scenes is again mapped on to the control of the city. Hugo was deliberately contrasting the hidden geographies of the poor in the 1840s with urban engineering later undertaken by Baron Haussman that built the grand boulevards for which Paris is now famous. The boulevards opened up the labyrinth of alleys to troops and police. Hugo was contrasting this open regular, state controlled geography with the earlier opaque and unknown city. The novel can thus be read as using the landscape to suggest a geography of knowledge, by the state about the potentially rebellious poor, and thus also a geography of state power. Lest this seems extreme, during the 1848 uprising some of the first things to be destroyed were street lamps – lamps that allowed the police to see what the poor were doing. In Paris street lighting was a police responsibility – mapping the geography of public light on to that of state surveillance.

Box 4.1

Light, power and planning

The depiction of light and dark, of an urban scene opaque to outside knowledge are powerful themes that tell us much about the culture of planning. So Alan Sillitoe, who read and reread *Les Misérables* as a child, echoes it talking of Nottingham. And such is not inappropriate if we look at the city of Nottingham's planning history. In the post-war period, it too was engaged in clearing 'chaotic' dense urban areas – where the poor lived – and in building new housing estates described as being light, airy and spacious (Daniels and Rycroft 1993). The parallels with planning and knowledge in a modern city are striking.

In detective stories we can see a different working of the themes of knowledge and control – suggesting less confidence in the ability to control urban life. Like *Les Misérables*, a recurring theme is how the city can be made intelligible, be made legible, to the forces of the state and, possibly, justice. The city is far more than a backdrop to the story:

> the spaces of detective fiction are always integral to the texts of detective
> fiction . . . the spaces of the genre are always 'productive' of the crime
> they contain and structure, forcing the detective to engage with the setting
> she/he inhabits in order to understand and therefore solve the crime . . .
> [to the detective] there is no stone in the street, no brick in the wall that is
> not actually a deliberate symbol – a message from some man, as much as
> if it were a telegram or postcard.
>
> (G.K. Chesterton 1902, cited in Schmid 1995: 245–6)

The detective is thus set up as an interpreter of urban life, rendering the spaces of the city legible. For instance, Sherlock Holmes ventures out to find knowledge about mysteries – often by going off into the darkest recesses of the city, into the opium dens and back ways. In the foggy London of Holmes, the central landscape features are the opaque mysterious goings-on in such hidden realms. Hidden because although Holmes goes into them – as a master of disguise – the reader rarely follows. The city is a riot of meanings, of significance, where the minutiae speak volumes to Holmes – but a city that cannot be read by us unaided. Holmes, however, can go anywhere, moving freely bringing order out of this chaos. The lights of Baker Street are beacons of hope and reason. Holmes is the embodiment of 'epistemological optimism', the hope and possibility that the city can be interpreted and understood through the power of reason.

In the stories of Raymond Chandler, with the detective character Philip Marlowe (transformed into film by Humphrey Bogart among others), we have different city and period. The Los Angeles of the pre- and post-war period forms the central feature of this genre of 'noir' fiction – so called due to the often nocturnal settings and the darkness of the cityscape. The city is full of dark spaces, the underworld is once more an unknown. If we focus on the Marlowe stories, Chandler's novels combine the psychological state of Philip Marlowe with the spatialisation of power in the contemporary American city (Schmid 1995). The two interact constantly, to suggest both something of a figurative geography and knowledge about the city. Chandler depicts a starkly divided geography, spaces of the rich – tending to be light and safe – are contrasted with those of the underworld in the dark city. Thus in *Farewell My Lovely*, Marlowe remarks of a rich home that there seems to be 'a special brand of sunshine, very quiet, put in noise-proof containers just for the upper classes'. It is indeed a landscape of division still remarked upon. The detective is the figure who not only traverses these spaces, but also reveals that despite their apparent separation they do connect – often through shady dealings by the rich. The detective is the one who knows this, giving rise to a cynical character. Here the city may be susceptible to interpretation but this is not always either a pleasant or comfortable thing.

Both these detectives are male. The experience of urban life and the interpretation of the city might be very different for women. Doris Lessing's *Four-Gated City*, suggests that far from being threatening the complexity of meanings and spaces in the city may emancipate women. This can be seen 'in the freedom and anonymity of the city [where] Martha recognizes the various personas and masks she has won in the past and she realizes that she can control them . . . She accepts herself as having many personas and being multilayered' (Sizemore 1984: 179). This city is not a two-dimensional mental map – say of edges and nodes as Lynch (1974) proposed – but a complex multi-dimensional map including the lives loves and histories of people. Martha, coming to London from an upbringing in the colonies learns about the city through the fragmented lives of urban characters. The great monuments of the imperial capital are displaced – not evoking imperial grandeur but instead meaning that 'all-kinds of half-buried, half-childish, myth bred emotions were being dragged to the surface: words have such power! Piccadilly Circus, Eros, Hub, Centre, London, England . . . each tapped underground rivers' (Lessing, cited in Sizemore 1984: 183).

The gendering of the city in novels and the gendering of knowledge about cities can be vitally important. A different landscape of Paris emerges in Emile Zola's *Au bonheur des dames* where the focus is on a new urban space of light and commerce – the early department store, the Bon Marché. At the end of the nineteenth century such stores were novel urban spaces – forming an imaginative geography of commodities and desires (see also Chapter 8). Zola describes these 'dream palaces', which promise so much and arouse so many desires for goods through their opulence, as creating worlds in themselves which are neither totally real nor totally illusory. The store is shown as a world of women – assistants and consumers – whose desires and wishes are readily apparent to the male store owner (Mouret), so that 'the double movement of limitless fantasies supported by strategic rational planning is figured in the description of Mouret's mistress' (Bowlby 1985: 72). In Zola's words this 'cathédrale du commerce moderne' is a feminised space controlled by masculine knowledge and desire. 'Mouret's sole wish was to conquer Woman. He wanted her to be queen in his establishment; he had built his temple to hold her at his mercy there'. Behind the counters the female staff work in a Darwinian jungle, struggling to survive, as nameless numbers living in the lofts.

The novel then speaks to us about a gendered geography of the city. Focusing on the urban space created through the department store it outlines a geography where rational knowledge and control, male power, economic prosperity and hardship, and gendered desires come together. This section has tried to suggest how looking at different novels can suggest complex and fascinating geographies, showing relationships between knowledge and power, knowledge and gender and the economy coming together in different ways. If we think about this we can also see these are social texts speaking about the contemporary hopes and fears for urban life.

riting modern urban experience

If we take Paris in the nineteenth century as a point of departure, we can see how the sensation of urban life is shown to be changing. Central to this is the concept of modernity, a 'structure of feeling' created through industrialisation. The expansion of cities meant that they were too large to know. To see what this means we can contrast the idea of a village to a city. Turn of the century urban theorists (such as Tonnies and Simmel) contrasted city life with a sense of village *community* (called *Gemeinschaft* in German) where everyone knows everyone else – their job, their history and their character – and the world is relatively predictable. Such an order is made problematic by strangers, people about whom nothing is known, about whom people have no preconceived ideas, no material on which to judge their likely actions. In modern cities, they suggested that for more and more people life stopped being so dominated by communities and became a world of strangers. The city is a constant series of contacts with people about whom very little is known, and who know little of you. Such is a shift to urban *society* (called *Gesellschaft* in German).

This frenetic bustle, Simmel (1990) argued, led to both excitement and loneliness in the crowd. The city comprised the dual features of *anomie*, that is, isolation amid the fleeting and fragmentary experiences of *Gesellschaft*, and also a vast increase in the stimuli and novel experiences to which an individual was exposed. The coping strategy he suggested was that the urban dweller rapidly became blasé to new events. In literature, a figure called the *flâneur* begins to be written about in mid-nineteenth-century Paris. This is a dedicated stroller, who has the leisure time to treat the frenetic motion and turbulent life of the city as a spectacle. The figure emerges as closely bound up with the first stirrings of urban journalism – appearing in the *feuilletons* of nineteenth-century Paris as a figure both observed and as a figure of the commentator. The *flâneur* became one of the folk-types of the modern city – feasting his eyes on the flow of new commodities in new shopping spaces (covered arcades and department stores), enjoying watching the truck and barter of the street. Note how I say 'he' for this figure is very often a male figure – the public arena was not regarded as an appropriate one in which bourgeois women could just idly wander. The male figure fascinated by commodities provides a contrast to Zola's work on women obsessed with commodities – but it also says something about changing urban spaces: the enclosed department store was designed to 'interiorise' the street, to make it into a private space, both under the control of one owner, but also an acceptable place for bourgeois women to shop.

There is an extensive range of writings on the *flâneur* and on the practice *flânerie*. In a prominent example, the *flâneur* was the figure used by Baudelaire in his poetry about Paris and to a certain extent reflects the artistic practice of the day. The figure was an anti-hero that went 'botanising on the asphalt', that is, he subjected urban life to the detached curiosity and categorisations normally reserved for the natural world. The figure emerges as a paradox in several ways: he is about 'leisured' time but watches the increasing velocity of modern life; he stands apart from the urban buying and selling, yet is fascinated by the gorgeous new displays; he inhabits a public space dominated by men, but watches the thousands of anonymous lower-class women who are the shopworkers, mistresses and streetwalkers of the artistic world. The *flâneur* watches the pace of modern life accelerate from his own dawdling pace – so that the apocryphal fashion was to take a lobster out walking in order to not go too fast. He embodies time becoming money – the *flâneur* can show his wealth by not hurrying, by wasting time – as the circulation of money and goods accelerates, his slowness forms an ever-clearer contrast. He does not need actually to buy the commodities, since consuming them visually provides satisfaction and displays wealth. From these practices we begin to gather a structure of feeling for 'modern' life, or indeed modernity. The phenomenon of a city of strangers produces alienation but also turns itself into a spectacle. From *flânerie* to department store, we can see the transformations of urban space. As the city was lit up with gas light, as glass-covered arcades opened and mass-produced good swelled the marketplace, the city itself became a spectacle of goods and events. This is not simply an architectural or economic shift; it changes the experience of the city.

Literature as a practice partakes of these changing experiences. The *flâneur* has strong autobiographical links with the experience of writers such as Flaubert and Baudelaire. Moreover, this comes through in the style of writing, in the city created through the written text. Thus far from treating literary works as things that simply portray or describe the city, a source of data, we must look at how they construct the city in different ways. As Brosseau puts it:

> Most geographers have seen the novel as a dead object, an inactive 'ready made source for the social sciences' that yields its information in an almost transparent fashion. Novels have been considered as geographical texts which can be combed for 'relevant' spatial elements in order to evaluate how good a geographer the novelist is.
>
> (1995: 91)

Instead we might look carefully at how the city is constructed in these novels – as we began to with detective fiction and Victor Hugo – in order to see how modernity is not just described but becomes part of the way of describing the city. Thus the work of Baudelaire is not only an account of the city, but the text itself seems to be a practice of *flânerie*, where 'the city becomes encounters "stumbled among like words"' (Robinson 1988: 193). His solitary wanderer moves among a multitude of people and encounters, but can never grasp the total city – the urban experience does not allow such a vantage point. The poet Flaubert equally wrote in a 'fugitive and momentary mode' (Robinson 1988: 201).

One of the significant shifts can then be seen in how literary forms deal with space and time – how the space of the city becomes fragmented, and how time is seen as increasing in velocity as the rhythm of urban life quickens. This can be seen coming into the twentieth century. In the nineteenth century the dominant mode of the novel was the narrative account, but in the twentieth new forms develop – such as the free-form recollection in Marcel Proust's *A la recherche du temps perdu*, where the account moves along a series of digressions, triggered by momentary experiences and the memories they trigger, giving an account of time that is far from a linear progression. At the same period, stream of consciousness novels emerge, notably with writers like James Joyce or Virginia Woolf, suggesting the inability to form a coherent narrative, because a narrative requires a grasp of the whole plot. These forms break up the time of narrative realism and problematise how the experience of modern life can be represented. Such a crisis in the way the city is represented occurs at the moment when telegraphy, the telephone and electricity were transforming communications and the scale of urban space. Stephen Kerns (1983) argues this technological shift underlay an acceleration of life that broke up ideas of stable vantage points from which to depict the city – not just in literature but in art with the decline of perspectival models in favour of Cubism. These accounts suggest that the acceleration of modern life causes problems for humans in understanding the world and making meaningful accounts. The existential crisis this sense of time accelerating provokes in literary styles is noted by Lukács:

> obviously if one looks on both social and personal life as pointless and sees reality revealed in the inevitable and wretched failure of the best human aspirations, then time, too, and its presentation must assume a new function . . . if life is pointless time must be seen as an independent and remorseless machine which flattens, levels and destroyed all personal plans and wishes, all singularity, personality itself.
>
> (Cited in Robinson 1988: 198)

This leads to a style that 'decapitates time' according to Sartre, leaves little room for the reasoned narrative of change, and thus the critique of change in reasoned narratives. The relationship of descriptions or representations of the world and the form of the account raises important issues not just about novels but about the most appropriate textual form for geographers to use in their own work. We might ask whether it is appropriate that geographers' accounts of cities should still cling themselves to a model of narrative realism.

Perhaps they can learn from novels such as *Manhattan Transfer* by Dos Pasos. Brosseau (1995) argues that the form of the novel is geared to the experience of life in twentieth-century New York, with a staccato pace that suggests the experience of a 'fragmented city'. There is no clear narrative to temporally locate events or cause and effect, instead there is a juxtaposition of scenes of poverty and wealth – the sharp contrasts that mark urban life but also show the lack of clear connections in that life. The novel implies these connections by thus writing on to the space of the city its uneven worlds of opportunity. The plurality of the city is shown where narrative lines relating to different places unexpectedly collide or cross-cut, enacting the rhythm of daily life in the form of the text. Reading the text becomes like walking on the sidewalk itself, not watching someone else do so. In this way the work goes beyond being a text on the city to being a fusion of urban experience and text itself. It stops being a single account and takes into itself the plurality of experiences in the city.

ummary

The text does not simply reflect an outside world. It is misleading to look at how 'accurately' or otherwise it corresponds to the world. This sort of naive approach misses the most useful and interesting elements of literary landscapes. Literary landscapes are best thought of as a combination of literature and landscape, not with literature as a separate lens or mirror reflecting or distorting an outside world. Equally literature does more than simply provide an emotional counterpart to an objective knowledge in geography. Rather literature offers ways of looking at the world that show a range of landscapes of taste, experience and knowledge. To say it is subjective is to miss a key point. It is a social product – indeed in circulating ideas it is a social process of signification. It is a social medium. The ideologies and beliefs of peoples and epochs both shape and are shaped by these texts. They shape what authors feel able or driven to say and how they say it. In this each text will draw on others – in part it is read in terms of conventions it either utilises or overturns. It is trying to speak to an audience and must thus engage with their expectations and concerns. It may change or challenge this but it must be comprehensible. Thus the intended readers mark their presence in what any author may be able to write.

To this end, literature is not a mirror held up to the world but part of a complex web of meanings. Any individual account will act in relation to other texts. Now these need not all be literary – they may be in other media (Chapter 5) or different types of literature (official accounts, promotional leaflets, or even academic works). Texts work to create networks of association between ideas in order to create ways of seeing the world. 'Realism' is one such constellation of links not a standard to judge a work by. Realism reflects one set of urban experiences – other literary styles may reflect different experiences. Here we may then also jump ahead to Chapter 11 and ask whether geographical accounts are so different from literature. Each is trying to open a particular way of understanding a landscape; each draws on other works; each draws on conventions of appropriate writing; each engages with the assumptions of its audience; each uses styles and rhetoric to provide a convincing vision. We should not see geography and literature as two different orders of knowledge (one imaginative and one factual) but rather as a field of textual genres, in order to highlight both the 'worldliness of literary texts and the imaginativeness of geographical texts' (Daniels and Rycroft 1993: 461).

Further reading

Abbeele, G. Van der (1991) *Travel as Metaphor: From Montaigne to Rousseau.* University of Minnesota Press, Minneapolis.

Cresswell, T. (1993) 'Mobility as Resistance: A Geographical Reading of Kerouac's "On the Road" ', *Trans. Inst. Br. Geogr.* (NS) 18: 249–62.

Frisby, D. (1985) *Fragments of Modernity.* Sage, London.

Jeans, D. (1979) 'Some Literary Examples of Humanistic Descriptions of Place', *Australian Geographer* 14 (4): 207–14.

Leed, E. (1991) *The Mind of the Traveller: From Gilgamesh to Global Tourism.* Basic Books, New York.

Pocock, D. (ed.) (1981) *Humanistic Geography and Literature.* Croom Helm, London.

Porteous, D. (1985) 'Literature and the Humanist Geographer', *Area* 17(2): 117–22.

Schmid, D. (1995) 'Imagining Safe Urban Space: The Contribution of Detective Fiction to Radical Geography', *Antipode* 27(3): 242–69.

Squier, S. M. (ed.) (1984) *Women Writers and the City.* University of Tennessee Press, Knoxsville.

Squire, S. (1988) 'Wordsworth and Lake District Tourism: Romantic Reshaping of Landscape', *Canadian Geographer* 32(3): 237–47.

Stallybrass, P. and White, A. (1986) *The Politics and Poetics of Transgression.* Methuen, London.

Tester, K. (1995) *The Flâneur.* Routledge, London.

Williams, R. (1973) *The City and Country.* Cambridge University Press, Cambridge.

Self and other
writing home, marking territory and writing space

- Relational models of identity
- Imperial literature
- Gendered landscapes
- Orientalism

This book started by noting how the diversity of cultures around the world formed one of the main stimulants to cultural geography. However, the study of the geography of cultures was deeply entwined with empire-building. This chapter will attempt to see how the popular dissemination of imperialist ideas shaped understandings of cultures and what legacy this has left for cultural geography. This is not to claim a primacy for ideas in motivating imperialism but to look at the mutual entanglement of the imperial and geographical imaginations and projects. It takes the term 'geography' at face value, in its etymological roots of 'writing the world', that is, inscribing meanings on the globe. This chapter not only explores how accounts of colonised peoples were shaped, but how these ideas reciprocally shaped Western identities. A key idea is that the identities of coloniser and colonised were relational – that is, the one depends on the other. Ideas of what it meant to be Western were shaped by ideas of what it meant to be non-Western. This chapter will look historically at these identities and will suggest that, although the formal trappings of empire may have ended, there may be a deep-seated and lingering legacy in terms of how Westerners understand the world.

This chapter starts with the beginnings of empire in the invasion and conquest of the Americas. It will then examine the relationships of Europe to the Orient and Africa. It will suggest that writing the 'foreign' helped construct a notion of the 'home' culture through a process of 'Othering', whereby the 'self' is defined in relation to the characteristics of an 'Other' culture. The materials will be drawn from the explorations and travel accounts of Western writers. The chapter will conclude by raising questions about how these processes formed the background for geographical studies.

Othering

Much recent work in cultural geography has been on the constitution of identities. These identities can be seen at individual, group and national levels, and are often formed by beliefs of common ancestry, or experience, giving rise to shared characteristics or traits. However, things are not this simple. For a start, very few people are the 'same' as others – everyone is different in some respects. The most we could say is that certain groups share certain things in common, so who is counted as part of a group or excluded from it will depend on which things are chosen as being significant. Looking around a lecture hall, very different ideas of groups sharing an identity would be formed if we used gender as a common factor, different again if we used sexuality, again by age, different again by income, different again by ethnicity, and so on. Belonging in a group depends on which of all the possible characteristics are chosen as 'defining' membership. The characteristics that have been treated as definitive vary over space and time with significant political consequences attached to deciding what defines belonging.

Some characteristics might be described as *elective* – you may choose to be left-wing or right-wing, you may choose indie music or rock. Others are *ascriptive* – our sex is generally given as is our colour. However, neither of these are really that clear-cut. The colour of a person's skin only assumes a significance when groups in society give it such importance; being genetically female does not entail an aptitude or desire for housework, but society may prescribe that role as appropriate. The meaning of even biological categories is given through social mechanisms – they do not have a natural or pre-ordained significance. (The relationship of 'factual categories' to cultural geography is examined at the end of this chapter and in Chapter 11.) Categories of identity are neither solely voluntary nor naturally given. Categorising people is a political process, where the stakes are often to define taken-for-granted natural, unquestionable categories. This chapter will suggest that it is impossible fully to think through how people can have an identity, that is, be defined by shared characteristics, without working out

Box 5.1

Relational identity

Identity can be defined as much by what we are *not* as by who we *are*. This is where geography often comes in since these 'us' and 'them' groups are often territorially delimited. We use a spatial shorthand to sum up characteristics of other groups – they are both defined by where they live and in turn define it. Linking with ideas of territoriality and attachment to place (Chapters 4 and 7), this chapter will explore how relationships over space become involved in defining group identities. It will be suggested that space is crucially involved in defining 'other' groups. A process often termed *Othering* through which identities are set up in an unequal relationship. The first group defines itself around a common feature (say *a*) and then defines all non-members as a residual (*not-a*). Obviously what is an elective identity for one group is not for the other. Furthermore, the tendency is to group around what are perceived as 'good' characteristics, thus whatever defines *a* will tend to be positively valued. Now assuming most people are a mix of good and bad points, this leaves a slight conundrum for the people *a* of what to do with their less desirable side. This chapter suggests the tendency has been to project the fears, 'the bad points', of a group on to outsiders. Thus part of belonging to a group is the projection of fears and dislikes on to other people. See also Chapter 10.

who is thus excluded – how identity is founded on differentiation. In simple terms, it is a situation of us and them. It is difficult to conceive of how we would define ourselves as a group ('us') in whatever manner without a contrasting 'them'.

The mapping of identity on to geography exposes the unequal relationships between groups and the importance of naming or being named, being the subject or the object of this process. Thus Richon (1996: 242) notes that terms like the 'Orient and the Occident are not just words, but names, proper names constructing identities which have become territories'. These territories have only been made apparent by a Western surveying gaze that constructs itself through looking at the Orient, while the 'Orient' exists only through that gaze. This chapter will suggest that this relationship leaves the subordinate group as 'objects' of a knowledge that denies them the right to shape their own identity and uses them as a 'negative pole', being the devalued or disliked elements, around whose exclusion the dominant group's sense of self can be organised. However, we should note that in projecting fears groups also tend to project forbidden desires on to outsiders, so it should not be surprising if fears and desires sometimes get mixed up in this process.

This can be seen when groups form identities by excluding what they fear – making it desirable because it is forbidden and unobtainable. In reality, people are not defined by single characteristics so throughout the chapter there will be points where there are struggles and shifts as people try to negotiate their position in all of this – linking often contradictory characteristics and positions that would label them as both part of one group and part of another.

Encountering America

Let us start this story with the European invasion and conquest of the Americas. Already we have glimpsed the crucial cultural links and changes that went on in the process of adapting to a 'New World' (Chapter 2), but here I want to focus on what those who stayed behind made of it. Few can doubt the enormous impact of the 'discovery' of the Americas on Europe. Suddenly the knowledge of the Ancients, the biblical orthodoxies, even the ongoing conflict with the East took on a new perspective. The shock of the discovery for the Europeans was considerable. How was something so unexpected to be assimilated, made sense of and understood? The invaders drew on prior accounts and ideas already afloat within their society in order to communicate about the lands and peoples they were conquering. Such images inevitably came through the situation of invasion, subordination and plunder. We might identify two *tropes* that became very influential in discussing the indigenous peoples.

Box 5.2

Tropes

These are ways of telling a story, through a particular format, a scenario or relationship of characters so that the pattern is repeated in different concrete situations with different contents. We might think of the westerns that follow the plot of a ranchers' way of life being threatened by 'sod-busting' homesteaders who intimidate them with hired guns till eventually one of them 'stands up' to the rancher and is killed after which all the others rally round. Or there are detective movies where the villain escapes on a technicality, forcing the cops to break the rules to get round their superiors. These plot-lines have appeared in umpteen different films and the outline remains the same, however much the place and characters change. See Chapters 4 and 6.

The first trope is the characterisation of the indigenous peoples as 'Noble
Savages'. That is, they are seen as simpler, purer people – indeed in those
religious times, as people from before the Fall. The Americas are, then, a
Garden of Eden which is polluted with regret by the fascinated
Europeans. In the second trope, the indigenous peoples are seen as lower
orders of humanity, indeed sometimes as separate species. They are
depicted as quite the opposite of Europeans: unclad, openly sexual,
illiterate, they form a contrasting foil that defines the virtues of Western
civilisation through its opposites. As Michel de Certeau (1988) looked
through tales of early explorers, he suggested the following pattern:

Western	**Americas**
clothed	naked
fashion	adornment
labour	leisure
ethics	pleasure
masculine	feminine
reason	emotion
culture	nature

Let us try and exemplify this by looking at how de Certeau begins his
book *The Writing of History* (1988) with the etching by Jan van der
Straet (1619) (Figure 5.1). In this picture the invader, Amerigo Vespucci,
is depicted standing in front of an unclad woman reclining in a
hammock. Commenting on this de Certeau suggests:

> Amerigo Vespucci the voyager arrives from the sea. A crusader standing
> erect, his body in armor, he bears the European weapons of meaning [a
> sextant, symbolising navigation, and the royal banner of Spain claiming
> the land]. Behind him are the vessels that will bring back to the European
> West the spoils of paradise. Before him is the Indian 'America', a nude
> woman reclining in her hammock, an unnamed presence of difference, a
> body which awakens within a space of exotic flora and fauna.
>
> (1988: xxv)

In this allegorical picture, the relationship of the two figures expresses
the contrasting signs of different groups. Europe is identified with
science and rationality (the sextant), it has its own name (Amerigo
Vespucci) and is going to claim and Name the Other – indeed with a
corruption of the invader's first name. The Americas are symbolised by a
naked female figure showing either innocence or sexuality – reclining

Figure 5.1 *Vespucci landing in America. Allegorical etching by Jan van den Straet for Americae decima pars by de Bry, 1619.*

and associated with leisure. The nakedness not only contrasts directly with the clothing of the invader but also implies a life of pleasure in contrast to the constraining armour. Europe and America are defined in terms of each other, but clearly not in an equal manner.

The feminisation of America cannot go unnoted. Later on the position of women as colonised and as coloniser will be discussed – showing how this simple division above was in practice more complicated. As the conquerors expounded on the fertility of the Americas – after all they were often writing to get sponsorship for their invasions – the language used was very often one filled with feminised descriptions – about fertility, fecundity but also eliding female with irrational and, importantly for invaders, subordinate. Pictures of naked women were erotically charged in periods where Western sexual mores were under fairly strict policing by the church. Ideas of sexual availability being natural were indeed used to excuse atrocities (and forbidden 'pleasures') by the European invaders. A lieutenant of Columbus wrote:

> While I was in the boat, I captured a very beautiful Carib woman, whom the said Lord Admiral [Columbus] gave to me. When I had taken her to my cabin she was naked – as was their custom. I was filled with a desire to take my pleasure with her and attempted to satisfy my desire. She was unwilling, and so treated me with her nails that I wished I had never begun. But – to cut a long story short – I then took a piece of rope and whipped her soundly, and she let forth such incredible screams that you would not have believed your ears. Eventually we came to such terms, I assure you, that you would have thought she had been brought up in a school for whores.
>
> (Cited in Cook 1995: 247)

In this extract the naked indigenous Carib woman is the object of Western lust and desire – a desire satisfied by raping the woman. At the end of the section, though, note how the author implies the kidnap, whipping, beating and violent rape 'revealed' her highly sexual nature that had somehow been hidden. A presupposed sexualised identity is used to justify colonial brutality.

Many writers have argued that the treatment of the land echoed that of the women. That Raleigh named parts of America Virginia served not only to promote an English cause but to claim they were 'untouched' – neatly denying the rights of prior inhabitants. Indeed the idea of a bountiful landscape was used to suggest that since the inhabitants did not labour (an arguable assertion anyway) they did not use the land and thus did not own it. Philosophers such as Locke argued that the central difference was between those who 'improve' the land and those who 'collect', with the former having a moral right and indeed imperative to take over land and increase its output. The argument was that such resources could not be allowed to be 'wasted' in the hands of the indigenous owners. That such output might not benefit the original inhabitants was not considered an obstacle. The instruments of science allowed the drawing of maps of empty spaces that could be divided and owned, sustaining a Western vision of the conquerers as agents of civilisation and the indigenous inhabitants as parts of the natural ecosystem. But if that empty landscape was part of colonising the Americas, it did not square with the visions of the East.

The mysterious East

The relationship of the East and Europe was a complex and often worrying issue. There was no possibility of claiming that the lands of the Near and Far East were blank spaces. They had already been filled with images and fears about the Orient for centuries. Rather than being emptied, the East was consigned to the past – as an ancient origin, not a current rival. The relationship was in contrasting forms of 'temporality' for Occident and Orient. The Occident defined itself as progressive, in the sense of making history and changing the world, while the East was defined as static and timeless. This pattern can be seen in thinkers from Hegel and Marx through to politicians such as Disraeli. Europe shapes the future, while the East can only experience repetitions. Thus Disraeli, the nineteenth-century British Prime Minister, in his novel *Tancred, or the New Crusade*, espouses an idea of a circular history in Persia; or, put another way, in the popular novel *Hajii Baba*, one character comments that one Shah simply undoes what the previous one did. Likewise, the Shah resists 'improvements' and medical advances such as vaccination. The West is thus defined as doing things to the East, the West defines itself as an agent of history, through its ability to act on the subordinate East. Thus in *Kim*, by Rudyard Kipling, set in the Indian Raj, it is the Western character who is associated with action while Eastern identities are symbolised through the lama's quietism and withdrawal from the world.

The 'imagined geography' of fear, repulsion and desire adds more dimensions to this map of the East. The idea of the Orient was in generous part constructed through characteristics the West wished to expel from its own self-image. The seemingly endless Western fascination with the 'harem' of the East illustrates how this institution became the crucible of a whole range of repulsions and desires (Figure 5.2). Westerners often expressed repugnance at polygamy, at the intrigue of the harem, at the idea of eunuchs and the decadence they felt it expressed. Western writers and artists returned to it as a subject over and over again. As a site of sexuality, often depicted with naked or semi-naked women (sometimes boys) it represented not just what was forbidden in Europe but also what was unobtainable about the East:

> The harem is a place which excludes any foreign look. Western
> representations of the harem are then the fulfilment of a wish to uncover
> what is hidden. If the depicted is orientalised, the depicting activity is
> obviously Western.
>
> (Richon 1996: 252)

Paintings that were almost photographic in their 'realism' were actually based on novels; under the guise of reporting the Orient they revealed a fascination and fantasy of male sexual dominance.

The sexually charged relationship to the East went further than this. In what James Donald calls one of the most 'dementedly racist' stories of the colonial era, Sax Rohmer wrote *The Mystery of Dr. Fu Manchu*. The story spawned sequels and a whole genre of films whose influence can be seen extending up to 1980s pictures such as *Black Rain* (in which Japanese criminals plot to destabilise the USA). In the original novel the Western narrator and man of science is confronted with the slave girl of Fu Manchu, Kâramanèh. The response says much about sexual desire and its repression:

> Her words struck a chord in my heart which sang with strange music,
> with music so barbaric that frankly I blushed to find in it harmony. Have I
> said that she was beautiful? It can convey no faint conception of her. With
> her pure, fair skin, eyes like the velvet darkness of the East, and red lips
> so tremulously near to mine, she was the most seductively lovely creature
> I had ever looked upon. In that electric moment my heart went out to
> every man who had bartered honour, country, all – for a woman's kiss . . .
> East and West may not intermingle. As a student of world policies, as a
> physician, I admitted, I could not deny, that truth . . . At the mere thought
> of a girl so deliciously beautiful in the brutal power of slavers, I found
> myself grinding my teeth – closing my eyes in a futile attempt to blot out
> the pictures called up.
>
> (Cited in Donald 1994: 176)

There is a readily apparent geography of a West desiring a feminised East

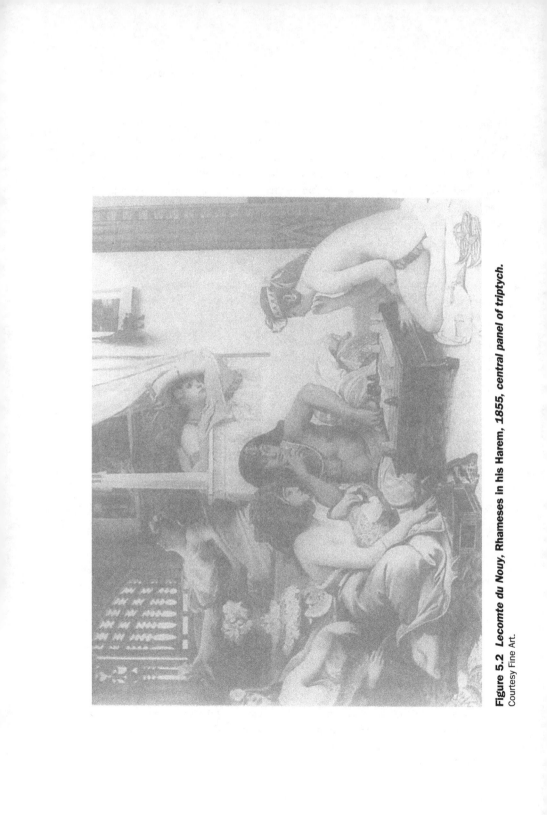

Figure 5.2 *Lecomte du Nouy, Rhameses in his Harem, 1855, central panel of triptych.*
Courtesy Fine Art.

– but as the object, not the subject, of desires. Equally there is a fixation on boundaries and definitions. The narrator 'knows' there is a line he cannot cross, a boundary that underlies much of the book. Fu Manchu is likened to a virus, infecting and creeping into the West, needing to be quarantined and shut out. The medical language is about purity – in the above case racial purity by resisting the sexuality of the East. The characteristics of East and West are contrasted vividly in the descriptions of the main characters:

> A breeze whispered through the leaves; a great wave of exotic perfume swept in from the open window towards a curtained doorway. It was breath from the East – that stretched out a yellow hand to the West. It was symbolic of the subtle, intangible power manifested in Dr Fu Manchu, as Nayland Smith – lean agile, bronzed with the suns of Burma – was symbolic of the clean British efficiency which sought to combat the insidious enemy.
>
> (Cited in Donald 1994: 185)

The value put on British men is directly related to making the Oriental Other the antonymical figure. We could move this account forward to look at the media coverage of the second Gulf War to see how Saddam was described as Mad Dog, labelled as insane and irrational. The whole description of the West's 'surgical strikes', its laser-guided 'smart' bombs echoes the language of Fu Manchu, a medicalised conflict with the West as a hyper-rational figure.

e Dark Continent

In the late nineteenth century there was a dramatic 'scramble for Africa' where European powers divided up the continent among themselves. The same logic of sexualising and feminising the continent can be seen at work. For instance in nineteenth-century art, say, Monet's *Olympia*, or further back to Gilray's cartoons, the presence of a black female servant signified not only sexuality but also deviant or uncontrolled sexuality – often appearing to indicate a fall from virtue or prostitution. Black male sexuality was seen as equally 'uncontrolled' but as a threat, through desires on white women. It is striking how in art 'the sexuality of the black, both male and female, becomes an icon for deviant sexuality in general . . . the black figure appears almost always paired with a white figure of the opposite sex' (Gilman 1985: 209). In these cases we can again see the flows of meanings and identity around issues of desire and

Figure 5.3 *Tourism advertisement for Morocco, 1994. The focus on a female figure, the interior scene and the description of enchantment and feasts raise some of the same issues as Figure 5.2.*

fear. Such art clearly displayed the racial ordering of identity. In Long's 1889 painting, *Babylonian Wedding Market*, the women are on display to be chosen by men according to their beauty. As they await this process the painting clearly shows a ranking of the women by whiteness. So whiteness symbolises beauty with the darker women portrayed as less comely. The whole picture is couched in horror at the barbaric process, yet also a fascination with this image of male dominance and female sexual availability. The colonial mentality was marked by a strong inclination to see black sexuality as a menace needing to be controlled.

This Othering of Africa expressed a need for control that was a projection of European inner fears where visual images portrayed the polar opposite of the European male. Thus to explain the roots of sexualised female images we have to search in the male observers. We can see Africa described as a fearful Dark Continent (in contrast with the civilised, white Europe that was slaughtering its way across it). It is the accounts of the West bringing light, civilising Africa, the accounts of missionaries flooding the continent with the light of reason and Christianity, that paint Africa so dark. Indeed 'Africa grew "dark" as Victorian explorers, missionaries, and scientists flooded it with light, because the light was refracted through an imperialist ideology that urged the abolition of "savage customs" in the name of civilization' (Brantlinger 1985: 166). Africa was occasionally portrayed as an Edenic world corrupted by European slavers in anti-slavery literature but the more general British attitude tended to see Africa as a centre of evil, possessed by demonic 'darkness', exemplified by slavery and cannibalism, which it was their duty to exorcise. The metaphor of bringing light is stressed in missionary accounts with titles like *Daybreak in the Dark Continent* and *Dawn in the Dark Continent*, and literary accounts such as the novel *Heart of Darkness* by Joseph Conrad. The popular geography of Africa related to Western desires and fears, the 'myth of the Dark Continent was thus a Victorian invention. As part of a larger discourse about empire, it was shaped by political and economic pressures and also by a psychology of blaming the victim through which Europeans projected many of their own darkest impulses onto Africans' (Brantlinger 1985: 198).

Such accounts centre on the European actor, the male protagonist in a feminised land. If we look at the immensely popular novels of Rider Haggard about southern Africa we can see this pattern very clearly. In his book *King Solomon's Mines* (1885), the landscape is both African and constantly feminised: for example Haggard writes 'these mountains . . . are shaped after the fashion of woman's breasts, and at times the mists and shadows beneath them take the form of a recumbent woman, veiled mysteriously in sleep' (cited in Low 1993: 197). Yet in this male-dominated landscape, women are both a figure of desire and fear. In Haggard's story *Nada and the Lily* (1892), which recasts the myth of two wild boys reared by wolves, one of the boys falls for a woman and is reproached by the other with the shame of desiring women, 'from whom ills flow like a river from a spring', depicting women as forces of instability and chaos. It is very much this troubled relationship which stands at the heart of the issue. Within Europe, Freud was busy developing psychoanalysis to deal with the 'problem of women' and describing the subconscious as the 'dark continent' of the mind. These links are important and they do not only impact on women or female characters, for these novels used the feminised landscape to create a stage where masculine characters could act.

These novels created spaces where male heroes could prove their worth with clear decisive actions. Haggard accused French and Russian novels of being morbid, the American writer Henry James of having feminine concerns in contrast to the manly pursuits of empire. There were hundreds of works set in the imperial frontier, be that Africa or Canada, that told stories of boys proving their manhood in feats of derring-do (Figure 5.4). This context is crucial since Haggard's stories fed into and were sustained by 'a cultural map of epics, travel narratives, exploration and the "boy's adventure"' set within imperialism abroad and growing militarism in public schools at home (Low 1993: 191). The contrast is with an urban order depicted as enclosing and stifling masculinity. Thus Haggard's Alan Quartermain flees urban society in order to develop his character. A project of defining an imperial masculinity links these adventure stories and the imagined and sexualised territories of empire.

Domesticating empire

Understandings of what it was to be 'foreign', of the non-Western world, keyed into ideas of what 'home' and 'homeland' meant. In part it could be read as the reverse of the colonies, the symbol of reason, justice and order. But given the imperial rivalries of the nineteenth century, home also became a ground for anxiety. And this anxiety was often expressed in racial terms, particularly in terms of the 'qualities' of the Anglo-Saxon race in Britain. They may seem odd now, but theories of racial destiny were then very common. Thus Robin Knox, in 1850, wrote that 'race is everything: literature, science, art, in a word civilization, depend on it', or Disraeli, in *Tancred, or the New Crusade*, looks to causes of empire:

> Is it what you call civilisation that makes England flourish? Is it the
> universal development of the faculties of man that has rendered an island

almost unknown to the ancients, the arbiter of the world? Clearly no. It is
her inhabitants that have done this; it is an affair of race. A Saxon race,
protected by insular position, has stamped its diligent and methodic
character on the century. And when a superior race, with a superior idea
to Work and Order, advances, its state will be progressive, and we shall,
perhaps follow the example of the [now] desolate countries. All is race,
there is no other truth.

(Quoted in Brantlinger 1993: 151)

Here once more we see both the importance given to racial categories
and how they are relational – the Anglo-Saxon is associated with work,
order and progress to the extent that other races, other cultures, are
marked by the absence of these virtues. Equally we can see that this also
creates the scope for racial anxieties, where the decay of a race is seen as
a real possibility and threat.

The claim of racial supriority in the face of subjugated and colonised
peoples also had the effect of assuaging fears at home. At a time of class
radicalism and antagonism at home, with the birth of the Union
movement and the Internationals in Europe, writers such as Kipling
could appeal to a white 'readership positioned as a racially homogeneous
and masculine community, unfissured by class allegiances' (Parry 1993:
223). Indeed Kipling, Haggard and others looked to empire as a cure for
alienation associated with a domestic and, crucially, urban working class.
Others used the context of British empire to unify the 'home nations',
bonding English, Scottish, Welsh and, in a more problematic way, Irish
identities under the overarching identity of British imperial pioneers.
Many movements seeking racial renewal, focused on the damaging
effects of urban environments as matters not only of social concern but
national survival.

king men

Patriots, such as Baden-Powell, were concerned that civilisation lead to
decadence, a 'softening' of men associated with urban life and 'moral
temptations'. So he turned to the frontier for a model of manhood that
would be equipped to rule the empire and defend it in a climate of
increasing imperial rivalry.

> *Scouting for Boys* expressed the middle-class values of the public school
> code and the Protestant work ethic. Its ideology was conservative and
> defensive, seeking to find in patriotism and imperialism the cure for an

Figure 5.4 *Cover of* Chums *magazine for boys, 11 June 1902.*

apparently disintegrating society. Its orientation was aggressively
masculine, its mission to save boys from the sapping habits of domestic
and urban life.

(Macdonald 1993: 8)

Baden-Powell recommended further reading, such as *The Race Life*,
paralleling the cycles of growth of individuals and cultures, alongside
manuals on military bridge-building. Such was not a solely English
concern; there were parallels in the German *Leibeskultur*, or body
culture, in fresh-air theory, in American Woodcraft movements, and host
of romantic movements appealing either to rural life or the frontier –
even as the age of imperial expansion came to a close. The stories of the
frontiersman in the USA, combating the militarily fearsome Amerindian,
find echoes in British imperial fiction about combating the Pathans in
India and the Zulus in southern Africa.

The practices and institutions in the imperial heartlands thus take up
some of the themes found in imperial fictions. And indeed juvenile
literature was often programmatic, depicting the frontier as a place where
men could prove themselves and learn the resilience they would need to
rule an empire. Ideas about masculinity and rationality can be seen
coming together clearly in the Royal Canadian Mounted Police –
portrayed as heroically masculine, mastering the untamed wilderness,
figures of order. The force became an emblem of male, imperial order
– an idea of manhood echoed in Baden-Powell appealing to Sherlock
Holmes which provides an interesting connection between the
landscapes of masculinity in crime novels and in imperial rhetoric (see
Chapter 4) .

aking women

If ideas of racial and cultural superiority scripted masculine roles, they
also produced roles for women in the imperial setting. Earlier sections
outlined the sexual imagery in popular ideas of the East with the idea of a
male imperialist. Parry (1993: 231–2) argues that in Kipling's novel, *The
Naulahka* (1892) 'The necklace of the title joins a sign of the East's
fabled wealth with a symbol of woman's body, and the narration of the
quest for the priceless and sacred jewel mimics a bellicose act that is both
an imperialist invasion and a sexual assault. A desolate landscape is
transformed into a meaningless social space, giving the West a moral
right to usurp its wasted resources.' India is a feminine landscape for the

male adventurer's actions. But what then of the women in India, or elsewhere?

The Western woman was a site of the contradictory relationships of gender and race. Her race was used to argue for superiority over the indigenous peoples, while her gender was more normally accorded a subordinate role. The terms used to make the colonised peoples sound subordinate were heavily gendered and normally used to justify and perpetuate the subordinate status of women. This has become an area which has attracted much interesting work on the varying situations for different women in different places – some found it gave them greater freedom than at home, others identified more with the colonised peoples, sometimes both occurred. For instance, Fanny Parkes's *Wanderings of a Pilgrim* (1850) provides little more than isolated picturesque vignettes of life in South Asia, giving no sense of agency to the colonised peoples, and in her entry into the female-only *zenana* she acts as the envoy for male fascination and eroticism. However, contradictions show through where she berates her ayha, her servant, for idleness because she goes back to sleep after dressing her mistress – ignoring the idleness this relationship suggests: white women were used to being dressed by a servant. She goes on to discuss her fifty-four servants, which seems to undermine her text's implication that it is white people who 'do' things, while also suggesting how class plays a vital role – she could not have afforded so many servants had she been in Europe. If men desired to rule, and portrayed natives as over-sexualised, then the sexual safety of female colonialists could be a real fear, especially as racial theories made men even more concerned about preserving blood purity by race and by class. So women could find themselves more hemmed in from contact. And upon their return to the imperial heartland some would change roles again: recipe books popularising Indian dishes in England were promoted by women who had strenuously avoided contact with the indigenous cultures while they were in India.

Geography and knowledge

Throughout this chapter there has been an emphasis on popular geographical knowledges and how these were shaped by, and went on to shape, relationships between cultures through the experience of imperialism. However, this did not happen independently of 'official' or academic geography. The cult of the explorer, as a geographer acquiring

knowledge about the unknown, was promoted by geographical
organisations. Bodies such as the Royal Geographical Society organised
(and still do organise) expeditions to go out and return to Britain with
geographical knowledge. At one level this is perhaps the romance that
first fired geography in this country, a sort of knowledge that Joseph
Conrad dubbed *Geography Militant*. In his childhood fantasies, the idea
of wide-open spaces for geographers to explore meant his 'imagination
could depict to itself there worthy, adventurous and devoted men,
nibbling at the edges . . . conquering a bit of truth here and a bit of truth
there.' But this romantic idea of geographical explorers was soured by
'the distasteful knowledge of the vilest scramble for loot that ever
disfigured the history of human conscience and geographical exploration'
(Conrad, cited in Low 1994: 195). We should think carefully about what
legacy there is for geography. For instance, this model of exploration has
left a heroic figure of the explorer as geographer *par excellence*. The idea
of the necessity of field experience in order to prove oneself a geographer
– a rite of passage – might also be argued to persist in the requirements
for fieldwork-based dissertations in hundreds of university curricula.

The cult of the explorer was fuelled by a jingoistic yellow press, with
expeditions bankrolled by newspapers to provide them with stories. Thus
Stanley's missions into Africa were closely linked with newspaper
circulation wars. The official geography was closely linked to some
popular media. Moreover, the image of the explorer was closely
modelled on that of the frontiersman. And we have seen the ways this
was set up to confirm a certain sort of masculinity and feminised the
peoples encountered. Finally, the knowledge was assembled into
institutions that operated closely with the bastions of science and the
military diplomatic system. The masculine edge of the explorer model
militated for a rationalist, empiricist form of knowledge. The detailed
empirical description of territories was not though a neutral scientific
pursuit. Not only was it underwritten by imperial interests, its very form
often served to perpetuate imperialist ideas. The fact-filled travelogues,
acted to reduce colonised peoples to so many vignettes – the impersonal
tone serving to obscure the relationships of oppression that made
'exploration' possible. The efforts that went into making data 'objective'
often served to cloak the violence of its creation or indeed the interests
behind it.

In thinking about studying cultural geography, we have to be aware that
claims for a universal, objective science have been deeply implicated in
racist and imperialist pasts. Claiming to speak from nowhere for

Box 5.3

'Objective' science and race

Many attempts to develop an 'objective' science about racial differences now
appear bizarre and would be comical if their repercussions were not so worrying.
For instance, there was a science of 'phalloplethysmography' which was that of
penis-measurement between races (Gill 1995: 39). In light of the gendered,
sexualised descriptions of colonised peoples, in light of the fears of white men
over the sexual safety of their women, all the 'objective' measurements tell us a
lot more about the concerns of the white scientist than anyone else. Apparently
factual accounts were clearly rooted in the fears and interests of Western men
working in science. Similar cases of objective knowledge riddled with implicit
ideologies about racial hierarchies can be found throughout debates over brain
size. The basic commitment of scientists to the idea of a hierarchy of races, can be
seen again and again to lead them to ask only questions that support that idea,
indeed to gather 'objective' data that support it. Indeed the argument was whether
the differences were so pronounced that races should be seen as different species.
Reams of scholarly work were committed to this, in Europe and the USA. Of the
latter Gould (1994: 93) makes the pertinent point that 'it is obviously not
accidental that a nation still practising slavery and expelling its aboriginal
inhabitants from their homelands should have provided a base for theories that
blacks and Indians are separate species, inferior to whites'.

everywhere has often meant speaking from the position of white,
Western man (see Chapter 11). As Parry argues (1993: 224) this
naturalises the ideas of the dominant culture as universal forms of
thought and accords its authorised representations the status of truth. The
Westerners name peoples and places, species and processes, according to
their own ideas of time and history – ideas which tend to leave other
cultures in subordinate roles. The geographical knowledge accumulated
through imperialism was marked out by a planetary consciousness. All
the world was put into typologies, hierarchies and subdivisions according
to a Western conceptual map:

> Within this type of knowledge, specimens are named by Europeans and
> extracted from their environment; in the process of naming them and
> setting them within a classificatory system, they are transformed from a
> chaos into an order that is European . . . Knowledge here is given the
> appearance of a simple neutral endeavour at an individual level, but in
> fact it is very much a part of imperialism; in this way scientific
> knowledge can present itself as free from the taint surrounding the
> commercial and political expansion it underwrote.
>
> (Mills 1995: 35)

It is important that we think how we as geographers are positioned when we do research. It is no longer adequate to claim some spurious neutrality; instead we need to think through how we relate to people we study, why are we asking these questions, not others, why are we studying them and not they studying us? Looking at the popular imaginations of empire should alert us to how ideas of rational inquiry bringing an orderly conception to the world have a long, and not always pleasant, history.

mary

This chapter looked at how imaginative geographies ascribed meanings to people and places through the construction of relational identities. This process has a historical geography bound up in the processes of empire – where divisions of a rational, progressive West and 'the rest' often legitimated Western and white supremacy. The chapter has tried to explore how this worked through a process of 'othering' whereby fears and desires from the dominant West were projected on to colonised peoples. This issue is picked up again in Chapter 10 which looks to ideas of national cultures in a post-colonial world and looks at the legacy of imperial ideas. The examples taken here show the subtle variations in ideologies about different areas of the globe and how they were perpetuated and supported by art, popular literature and social movements. It is important to note that this process is thus not simply about how 'they' were defined in negative terms but how those terms are closely bound up in 'our' Western self-definition. It is not only the Third World that has to decolonise and replace these ideas; the West needs to think what a post-colonial era means to itself. Finally the chapter has hinted how cultural geographies cannot be seen as standing apart from this process. The scientific study of race and culture has been a part of imperial processes, all the while protesting its objectivity. We thus need to think carefully about how we study these issues – a theme taken up in Chapter 11.

her reading

Blunt, A. and Rose, G. (eds) (1995) *Writing Women and Space: Colonial and Postcolonial Geographies*. Guilford Press, New York.

Brantlinger, P. (1993) *Rule of Darkness: British Literature and Imperialism, 1830–1914*. Cornell University Press, Ithaca.

Gill, A. (1994) *Ruling Passions: Sex, Race and Empire*. BBC Books, London.

Macdonald, R. (1993) *Sons of the Empire: The Frontier and the Boy Scout Movement, 1890–1914*. University of Toronto Press, Toronto.

McLintock, A. (1995) *Imperial Leather*, Routledge, London.

Mills, C. (1996) 'Gender and Colonial Space', *Gender, Place and Culture* 3(2): 125–47.

O'Tuathail, G. (1996) *Critical Geopolitics: The Politics of Writing Global Space*. Routledge, London (esp. ch. 3).

Parry, B. (1983) *Conrad and Imperialism*. Macmillan, London.

Phillips, R. (1996) *Mapping Men and Empire*, Routledge, London.

Riffenburgh, B. (1993) *The Myth of the Explorer*. Oxford University Press, Oxford.

Said, E. (1993) *Imperialism and Culture*. Vintage, London.

Smith, N. and Godlewska, A. (eds) (1994) *Geography and Empire*. Blackwell, Oxford.

Sullivan, Z. (1993) *Narratives of Empire: The Fictions of Rudyard Kipling*. Cambridge University Press: Cambridge.

Multiply mediated environments

film, TV and music

- Film and the rhythm of city life
- Creating spaces through media
- Aural landscapes: musical spaces
- Places and flows

The two previous chapters explored the way landscapes are depicted in literature. Literature is, however, just one creative 'media' through which cultural ideas are produced and reproduced. This chapter examines what geographers may gain from looking at other media – visual and audio. As with literature the first pass geographers made was to look at these as sources presenting landscape imagery, but, as with literature, we can look at these forms as creating geographies in a much stronger sense – actively shaping interactions in and with places according to various cultural norms. Moreover, these media 'intrude' into daily life, indeed given their prevalence they may be said to create landscapes in which the consumer is immersed, becoming part of the lifeworld of the consumer. These media are not separate from daily life; they are not an adjunct to human experience; increasingly they comprise the terms of the everyday world. These media are examined according to the themes developed so far – the creation of landscapes, the patterning of acceptable behaviour through the use of space and the relationships of mobility, modern life and the city. These media also highlight questions about the relationship of experience and media in the modern world.

m and the cityscape

Film can have obvious connections to literature – most clearly where films are made of books – raising issues similar to those in the last two chapters. This section will look a little wider by choosing particular genres – the urban detective movie, films which take the city as their theme, and the road movie – and asking what geographies they create. Inevitably, other genres of films can be used to identify different points, but through studying these limited selections I hope readers will develop the skills to think through other genres.

Chapter 4 outlined the way detective fictions in literature offer insights into the urban experience. However, Raymond Chandler's stories are perhaps more familiar through their filmed rather than written versions. The last chapter outlined how these particular stories constructed a landscape separating spaces of light and dark in the city. The urban realm is constructed through this contrast of honest dealings and the murky underworld. This is even more apparent in the films, where the city becomes a realm of dark spaces, through which the sense of danger is visually constructed. Such spaces are set off in stark contrast from the spaces of the well-to-do, with their well-lit and quiet arrangements. But equally these spaces seem all the more insecure, interspliced with the dangerous spaces of the city that at any moment threaten to destroy the harmony. The Los Angeles of these films is not the surfer paradise nor the sun-filled, expansive California of so much TV; instead the city is stitched together out of the dark underworld – the seedy underpinnings of society are clearly charted. The social divisions of urban life are mapped out in these light and dark spaces.

The city is as much an actor in the accounts as the characters. We can see this explicitly in films that took modern life itself as a focus. A film such as *Berlin: Symphony of a City*, a celebrated avant-garde film taking a 'slice of life' from the city, made by Walter Ruttman in 1927. Of course, just because it depicts the life of the city of Berlin does not make it a neutral reporter. *Berlin* is an artistic creation – although it is documenting life in the city, it deploys aesthetic tactics, to convey truths about the city. It actively 'writes' the city in the choice of scenes, the camera angles, the editing and content of the film. What then were the truths *Berlin* conveyed to its audiences about urban experience? First, we have to rediscover the wonder that audiences felt at the spaces created through film in those early days of cinema. Film allowed new spaces of observation, it created impossible visions that contrasted with previous ways of seeing the city. The film did not follow the conventions of a panorama – where the city is laid out as a visible whole seen from an elevated or aerial vantage point. Likewise the city was not experienced through the visual conventions established in the nineteenth century, in the diorama – where the static *mise-en-scene* has action unfolded within it. Instead the film took the experience of moving diorama and travel by mechanised transport where the world flowed past the window. Writers of the period, such as Georg Simmel, noted how the city opened up to an ever-increasing bombardment of seemingly disconnected sights and stimuli. Film offered a way of capturing this sensation by linking the diegetic (that is, on-screen) spaces of film in different ways to the plot – changing the connection of story, causality and space. Thus film-makers such as Sergei Eisenstein and D.W. Griffith shocked audiences with the 'cut away shot'; that is, filming one event then cutting away to show the next event in the story, but one occurring at another place, either at the same time or much later – there was no linear connection of the space and times shown. Events in two different places could be intercut – creating a physically 'impossible' viewing position, but emphasising the simultaneity, complexity and disjointed experience of urban events. Films, like the new literary modes in Chapter 4, helped shatter the unitary experience of place, bringing different spaces together to reveal new patterns of modern life. We can see film as suggesting the shattering of previous ways of living in time and space. At the same time Cubist art was challenging the traditional idea of perspective. Both these visual mediums are suggesting a changed experience of urban life. The city was not able to be mapped, or thought of, as if from one omniscient viewpoint that ordered all the relationships between spaces. Instead the spaces depicted in the film link to each other in more complex ways.

Ruttman's film of Berlin can be seen in this context. Berlin had emerged as the metropolis of Germany, the centre of electrification and rapid social change. It was characterised as a city of 'asphalt' to note the dominance of motor cars and commercial outlets. A city then of flows (of cars and electricity), of light (domestic and commercial, for living and display), of transient and unstable forms (both political and cultural). Amid this Ruttman's film:

> offers endless illustrations of the acceleration of life patterns and deindividualization brought about by work mechanisation, the emergence of a full-blown consumer society and its cult of distraction, and finally, the pure sensation of speed, at the workplace, in communication and transportation networks.
>
> (Natter 1993: 215)

The film charts the flows of people, energy and material around the city through intercutting and connecting shots, so before any human is shown 'the city is rendered as a physical plant, multilayered horizons consisting of sewer systems, and facilities generating steam power, heat and electricity' (Natter 1993: 217). Humans are also shown through movements and circulations. The film crosscuts between different classes travelling on foot, by bicycle, train and car. This emphasis on flows, on the interconnecting spaces of the city spatialises the city, fragmenting the experience of place. The effect of the film is not to present one place and one meaning, but to show the countless different connections that give countless meanings to each place as it is connected to others. The city of *Berlin* offers a plurality of meanings, networks and constellations of associations between spaces and flows rather than a single *genius loci*. At the time this was criticised by Siegfried Kracauer, for dehumanising the city by reducing its inhabitants to disconnected actions in fragmented spaces.

The relationship of ideas of humanity to the potentially alienating world of the modern city can be seen as a prominent theme in Fritz Lang's film *Metropolis* (1926). There are three crucial ideas to draw out of this film for our present discussion. The first is the vision of a futuristic city, the second is the alienation and fragmentation of human life almost literally buried under this vision, and finally the dissemination and wider connections of the film. Lang drew his vision of a future 'metropolis' after a visit to Manhattan, where the towering skyline had a profound effect on him. His *Metropolis* is a place of soaring towers, hundreds of storeys in height, connected by aerial bridges on which vehicles glide and between which personal aircraft fly (Figure 6.1). In establishing shots, the buildings soar from ground level, where there are specks of activity, to cloud-topped heights. There is a clear vision of progress here, and a clear geography to it. Lang, the German film-maker, had seen the future and it was American. More than that, this was a vision of the triumphant capitalist city, with ever-increasing wealth literally piling up higher and higher. It is a also a vision of technology transforming modern life, shaping new ways of living. In this sense it presents a purified version of Ruttman's city. Lang's *Metropolis* is the triumph of technical innovation in totally rewriting life, not merely accelerating it, not simply connecting suburbs and work in new ways, but transforming and creating new urban spaces and modes of living – in not very pleasant ways.

Underneath this tale of capital triumphant, is the labour required to sustain the modern industrial city. In chthonic spaces, in a buried world, below the city, Lang depicts the remorseless dehumanisation of the workers. Here life is lived according to the mechanical time of the hooter and whistle. Labourers all leave their homes and shuffle in line like

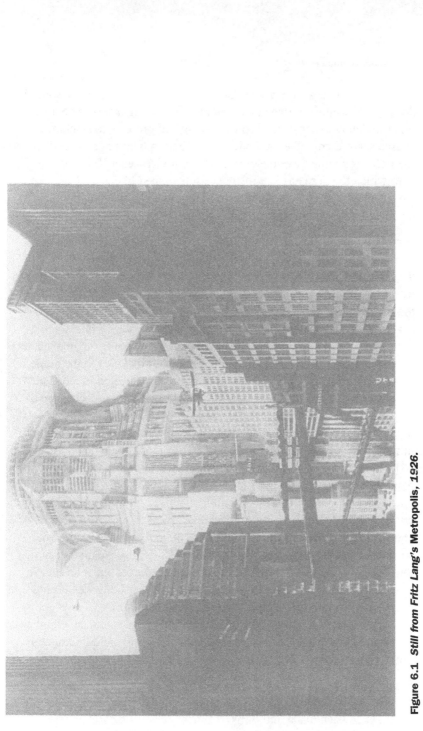

Figure 6.1 *Still from Fritz Lang's Metropolis, 1926.*

automatons to their workplaces. The main character is seen in a soul destroying struggle to regulate a machine, wrestling levers to control it till completely exhausted. Labour is totally subordinate to machines, people serve them rather than vice versa, and in their underground workplaces people serve the city rather than vice versa. The machine with which the hero has to wrestle is itself shaped like a clock – so he struggles to control the tempo of a life regulated by machines and mechanical rather than human time.

What are the wider echoes of this film? Well, in the same period, the architect Le Corbusier was seriously designing urban projects, such as the Radiant City, that were premised on totally expunging inefficient and irrational urban spaces that had developed over history and replacing them with cities of tower blocks. The project was to reshape domestic spaces as efficient 'machines for living', in a view where function was seen as all important. Just as industry looked at stripping out unnecessary parts or actions, functionalism sought to create efficiency in the home by reducing form to function. In practical terms this resulted in the idea of the tower block, as mass housing – like mass production – designed to supply an expanding modern society. Against the uncontrolled and unknowable spaces of the industrial city, as in Victor Hugo's account of urban life (Chapter 4), these modernist cities would be ordered, planned, rational and functional. The very term 'Radiant City' suggests on the one hand light, in contrast to the darkness of smokey nineteenth-century cities, and the idea of pattern and thus planning. As such we can see the connections of Lang's vision to ideas that animated post-war planning and reconstruction. The modern city was to be planned to avoid what were seen as the faults of the nineteenth-century city. We might say that the plans and the film are part of the same 'discourse' about urban life. They may be coming from very different quarters – Lang was far more sceptical of urban trends – but each mobilises similar ideas and concepts. Thus if we look at promotional films for urban redevelopment we find echoes of the move to make the city rational and functionally efficient.

Lang's film is a criticism of this vision of the ordered future. Along with writings such as Orwell's *1984*, or Huxley's *Brave New World* (both of which have been made into films), Lang suggests the costs of the 'Metropolis', in terms of dehumanising and subordinating people to the machine, are too high. Indeed his film ends in a revolt by the subterranean workers. Far from turning out to be a Utopia, a dream of perfection, *Metropolis* is in fact a *dystopia* – a nightmare. If we track forwards to the 1980s, many geographers saw this genre of film taken further by films such as Ridley Scott's futuristic *Bladerunner* (1984). Again there is a geography that animates this – a move from New York, as a model for the future, to Los Angeles. The film starts with a series of shots following aerial vehicles as they sweep among dark and threatening towers of buildings, interspersed with the sharp illumination of huge neon adverts. The dark city interspersed with these islands of commercial brightness is divided between the corporate headquarters and the teeming masses on the polluted street, speaking a street argot, full of seedy venues. The street is a threatening realm of disorder, and the corporate towers are fortresses against this. And travelling between the towers and the decaying street levels is the detective brought back to the police to negotiate these worlds. In Los Angeles many writers comment on the social divisions and conflicts in this poly-ethnic city, where the well-off pay for security against the poor. There is an urban geography of fragmentation and active spatial division. Far from the vision of working masses, there is what some would call an underclass – excluded from the (legitimate) economy. Yet it is also home of some of the most glamorous places and highest living standards on the planet. In *Bladerunner* we find the noir vision again, emphasising these divisions in terms of light and dark.

Such a dystopian vision of the city is by no means the be all and end all of looking at films. If Los Angeles produced a dystopian vision, Berlin of the 1980s gave rise to Wim Wenders's *Wings of Desire*. In this film mobility is again a key theme, yet this time restricted to insubstantial Angels who use the impossible vantage points of film to penetrate the spaces of everyday people. Floating across the city they listen to the alienation and solitude of people in the city, hearing not only speech but thoughts. The weightless mobility of the angels forms a contrast to the segregated spaces of flats and homes, the rooted repetitive life of ordinary people, the isolation and emotional divisions this causes. But to return to Ruttman's vision of a mobile people we might see this reflected in writings about Los Angeles, such as Joan Didion's:

> To understand what was going on it was perhaps necessary to have participated in the freeway experience, which is the only secular form of communion Los Angeles has. Mere driving on the freeway is in no way the same as participating in it. Anyone can 'drive' on the freeway, and many people with no vocation for it do, hesitating here and resisting there, losing the rhythm of the lane change . . . Actual participation requires a total surrender, a concentration so intense as to seem a narcosis,

> a rapture-of-the-freeway. The mind goes clean. The rhythm takes over. A
> distortion of time occurs.
>
> (Didion 1979: 83)

This seems more a return to the experience of the city as mobility
presaged in Ruttman's *Berlin*. In this rhapsodic account Didion is striving
to convey a sense of the speed and pace of life in Los Angeles, where
mobility has become the norm. This has come to provide a sense of air-
conditioned cruising, 'a community of the freeway' which is perhaps the
middle-class space for Los Angeles. From the suburbs to the city, it
bypasses the inner-city poverty – air-conditioned cruise-controlled
capsules taking their owners from home to work. Of course the suburbs
themselves may not be always seen as ideal. The car can symbolise
escape from claustrophobic suburban life as much as being an integral
part of the commuter city. Never one for understatement, Hunter S.
Thompson puts it thus:

> Every now and then when your life gets complicated and the weasels start
> closing in, the only real cure is to load up on heinous chemicals and then
> drive like a bastard from Hollywood to Las Vegas. To relax, sit as it were,
> in the womb of the desert sun.
>
> (Cited in Eyerman and Löfgren 1995: 53)

Thompson's novels may be extreme but the sense of escape through
mobility, and especially the idea of mobility in America, is worth
exploring. The genre of road movies has repeatedly reworked these
themes. These movies may have other elements in them, such as in the
film of Steinbeck's *Grapes of Wrath* the environmental and human
tragedy of the 1930s dustbowl in America is the main theme, but the
impoverished family end up having to become migrants, taking to the
road. And not just any road but Route 66, going West – which has
resonances in the whole myth of the frontier and colonising America. In
other films, the cause of flight may well be less tragic, instead it may be
an escape, from the claustrophobic petit-bourgeois suburbs, from an
intolerant 'normality'. Often the protagonists are out to, or are forced to,
find themselves (just as in the classic travel stories in Chapter 4). Other
times the story may be more disenchanted, such as in the 1969 *Easy
Rider* whose opening motif is 'A man went looking for America and
couldn't find it anywhere'.

We have, then, a particular configuration of space and time being used to
show the real America – or indeed its demise. As with Beat poetry, this is
also a dream about male flight from domesticity – linking man, machine

and mobility in a potent combination. Indeed only in the last ten years have road movies such as *Thelma and Louise* challenged this basic foundation. However, it is also important to think this through in terms of a geography of distribution and dissemination. For instance the road movies of Wim Wenders use the open spaces of America both as a mentality and a physical setting. For a European director, the whole possibility of driving in this way becomes Americana – a fragment designed to convey to the European audience something about America specifically.

sic and geography

Geography has often been dominated by visual material – from maps to films. It is important here to bring this up against the much less developed study of geography and music. One might wonder what geography can say about music and vice versa. This section will first look at how music may express similar relationships of mobility and space as seen in literature and film. Second, there are the spaces of ethno-musicology – that is, looking at how particular music is associated with particular places (developed further in Chapter 10). Finally, a *sonoric landscape* will be discussed in terms of practices of listening to music and how this orders places.

sic and mobility

Songs have long played an equal role in the mythology of the 'road', and mobility more generally, as novels or films. Many Southern blues songs took the refrain of the train going north as a symbolic route to the North – away from the racial segregation of the South. The folk songs of Woody Guthrie are often accounts of being a hobo, riding trains, migrating around the USA in the 1930s depressions. Indeed, one of his semi-autobiographical songs is 'Hard Travellin''. Lyrically his songs claim to speak for the poor forced to travel, but also lend a positive note to mobility – personally, once again Guthrie travelled away from domestic ties and also as an escape from poverty and drought. To take an example Guthrie's 'Oregon Trail' begins as follows:

> I've been grubbin' on a little farm
> on a flat and windy plain,
> Yes I've been-a listening to the hungry cattle music ball,

> I'm goin' to pack my wife and kids,
> I'm goin to hit that Western road,
> Cos I'm goin' to hit that Oregon Trail this comin' fall,
> goin' to hit that Oregon Trail this comin' fall,
> goin' to hit that Oregon Trail this comin' fall,
> Where the good rains fall aplenty
> and the crops and orchards grow,
> I'm goin' to hit that Oregon Trail this comin' fall.

This conveys aptly both the desperation of the times and the meaning that mobility has in the USA: the chance to start afresh, to make yourself over. This is a mythical geography, drawing on tales of pioneering and homesteading in the foundation of the nation, but it is as well to remember that Western countries are saturated with such myths. We might also see the gendered construction in the way the male protagonist of the song packs his wife and kids in much the same way as he later grabs one of his pigs by the tail and packs it off on 'that Oregon Trail'. The celebration of the highway stretches forward through Guthrie's influence on Bob Dylan, who in turn reproduced the fixation with the highway as a symbol for America – in albums such as 'Highway 59 Revisited' and songs such as 'Desolation Row' and 'Highway 61'.

People and their music

That these hymns to the highway are the myths and music of a particular landscape, and part of the Americana used by Wenders, can be illustrated by the satirical comment of the British 'urban poet' Billy Bragg whose answer to 'Get your kicks on Route 66' was the bathetic lyric 'go motoring on the A13'. As Leyshon, Matless and Revill dryly observe, 'the glamour of the transcontinental highway doesn't quite work in Essex' (1995: 430). This should remind us that we can read many European panics over popular music and culture in general as fears of 'Americanisation'. The geography of music has often contrasted local and universal, rooted and rootless. It has thus been led into a search for regional folk types – the mapping of styles and influences. This risks becoming a quest for the last remnants of such music as the electronic transmission and circulation of music has gathered pace. It has thus often become a quest to find the 'authentic' local style and songs.

At this level the geography is somewhat too simple. For much 'folk' music is an invention of the very people looking to 'recover' an authentic folk practice. Thus in England in the early twentieth century there were attempts to recover folk music (and dance) before it disappeared. The collectors, from the urban intelligentsia, went out to save the music from the folk. What they found were various fragments from which they attempted to reconstruct a single original template or master version. Such presumes that there was such a singular original. Many would now argue that they took a fluid, ever-changing range of practices and invented one 'authentic' origin to suit their own beliefs about folk music. Music could then be connected with feelings of belonging, and used to promote the idea of particular regional identities. Thus in England the model of an original set of songs which had been corrupted, or lost, or only remained in fragments, parallels social ideologies that England was being corrupted by urbanisation. Such movements might then become closely bound up in movements to find a national music, so we might chart the way Vaughan Williams tried to create an impression of the English landscape through using tones picked from 'folk music'. The parallels with literature are striking if we look at the poem *Ossian*, which was 'found' as a Gaelic epic poem. In fact it was 'reconstructed' from 'remaining' fragments by careful scholarship – on exactly the same premise as folk music. Thus interpreters knew that the classic cultures had epic poem cycles, they thus assumed what they heard were the corrupted remains of one. It is now widely doubted such an epic ever existed, but nationalist feeling dictated that all great cultures had pre-written epic poems. (See Chapter 10 for a discussion of such invented traditions.)

This treatment of music as embedded in a locale differs starkly from classical music where traces of the local have been gradually removed. Classical music has tended to treat itself as a universal, neutral standard – folk and ethnic music are measured as deviations from it. While notable developments occurred in specific places and at specific times, it is suggested the qualities of the music transcend this. Rather like the model of classical science, music became marked by its reproducibility. And just as in science this entailed the spread of particular spaces of controlled conditions and techniques – laboratories – in classical music there is the spread of the concert halls and particular practices of listening.

Landscapes of listeners

The geography of music may also be traced through the spaces of listening and performance, the creation of what might be called sonoric landscapes. Thus in paintings of the seventeenth century we see music associated with spaces of contemplation, social etiquette and the gentry – so in a 1658 painting by Van Schoor of the van Tilborough Nassau mansion, the music forms part of a scene of a walled garden. As noted in Chapter 3, this serves to re-emphasise the separation of a genteel private space separated from the surrounding world. It is not necessary to think only of historical examples of spaces of listening and performance. The concerts of k.d. lang have been suggested to form a particular space, where women can get together without an assumed heterosexual norm, a transgressive space that breaks conventional boundaries and the normally heterosexual medium of country music is subverted. We might make this point more broadly to suggest the meaning of music to the audience is context-dependent; so in these spaces, where lesbian women are the majority, they can appropriate this music.

Equally transgressions can be resisted. Thus in the 1930s there was considerable friction over the intrusion of urban, recorded music into the countryside. Campaigners for rural England, saw trippers bringing urban music into the countryside as the least desirable sort of visitor. To think of a simple example, in one of Arthur Ransome's children's stories *Coot Club*, the heroes, local sailing children, are set against trippers. The trippers ignorantly threaten the nest of a coot, symbolising their irresponsibility. These trippers, in a motorised boat from the seaside town of Yarmouth are known as the 'Hullabaloos' due to their gramophone. The 'moral geography' of middle-class rural, sailing children conserving nature is set against the motorised, noisy (for which read urban music) trippers from the popular resort on the coast. The conflicting geographies of taste and class are mapped out across the Broads of Norfolk.

In contemporary Britain this has echoes in the policing of raves – of predominantly young urban folk congregating in rural areas. The Criminal Justice Act of 1994 gave wide powers to the police to stop cars, to create exclusion areas and, specifically, to ban the playing of music with a 'repetitive beat'. So when thinking of the geography of music we need to think of the spaces created. We might then begin looking for transient 'spaces of affect' created through shared reactions to music. Spaces of dance and listening can create affective, emotional communities – spaces created in rural England; listening to a music developed through house music, the gay scene, Europop, disco, and thus back to the New York disco scene; or bhangra beat music drawing on an Asian culture, that has translated rap and disco with South Asian styles to create a wholly new form. (The spatial relationships in such music will be explored in Chapter 10.) It is not just a matter of associating bands with places, or even observational lyrics about places, but also the way music forms spaces for people – say the festival scene from Glastonbury to Tribal Gathering. For now we might note the creation of spaces of common belonging in youth culture. Where what the French sociologist Maffesoli (1995) terms 'neo-tribes' can gather – finding a community and a shared identity through dance spaces. Music of different sorts is opening up spaces of sociality where groups of people can come together in particular ways with different social norms – about gender and sexual roles, about alcohol and other drugs, about night and day. The fleeting, fragmented geography of club cultures presents a newly emerging and constantly changing sonoric landscape.

ographies of viewing

The discussion of mediated environments thus leads away from simply geographic content for various media, through to the spaces media create, and finally the spaces in which the media is used. It is in this last light that we might consider the geographies of television. It would of course be entirely possible to replay the arguments presented above about films. The difference about television is the time and space within which it is watched. Television produces a paradoxical geography of global flows and local embedded viewing. This section will try to open up some of the possibilities this creates, beginning with a discussion on how the 'sitting room' and the global articulate. It will then pose some questions regarding the impact of this in terms of social fragmentation and possible concentrations of power. In response to these concerns the final section will outline two counter-tendencies of television-viewing as creating communities.

obal sitting rooms

It is now over thirty years since Marshall McLuhan became a media prophet by declaring that what made a TV-based society remarkable was not the content of the programmes but their mode of delivery. The medium is the message. What he noted was the instantaneity and ubiquity of information provided by television. The speed and quantity were similar to how news might spread in a small community yet the range of TV meant that this could include the whole world. Hence he suggested we were now entering the era of a global village. It is this interpretation that TV leads to an enhanced, if not entirely new, global or planetary consciousness that has been drawn on to think about recent responses to global crises. The 1980s saw the rise of globally oriented charity events such as Live Aid. These used television to bring distant problems into the sitting rooms of people in the West. They used not only the speed of television transmission, but the immediacy of its images and its access into the private home spaces of viewers to bring the issues home. TV seemed to offer a connection of care and responsibility between those physically at far ends of the planet.

The space of flows in the new media have not only brought distant places together but have counterposed traditional forms of power. In the case of the fall of the Berlin Wall, we see how, no matter how formidable the forces controlling territory, they were susceptible to the flows of media in the airwaves. Similar issues were raised with the transmission of events in Tiananmen Square and the student protests there in 1989. Given that all the media coverage did not ensure the students' victory we need to think seriously about the balance of flows and territorial control in specific cases. Some commentators have speculated that the global media are part of an ongoing evolution in the forms of power in society. Where the actors used to compete through the control of territorial space, there has been a gradual shift to flows as more important. It is perhaps not surprising that the same period has seen a number of panics over maintaining boundaries – both by states and individuals. State regulators have wrestled with the possibilities of global media evading their control at the same time as parents have wrestled with issues of controlling what children watch. The maintenance of boundaries – be they from political or moral geographies – in the face of the de-territorialising flows of media seems one of the issues brought to the fore in geographies of television.

lenation, manipulation and fragmentation

There is a less benign view of the geographies created through television. One starting point is the very idea of the flow of information. It is easy to point to dramatic events but what is the bulk of this flow? If we are entering an 'information age' as has been prophesied what is notable is a lack of discrimination in what this information might be about. A cynic might follow Bruce Springsteen's bleak lyric that there are 'fifty-seven channels and nothing's on.' Not only might we query whether the quality of lives will be improved by this but we might think about the effects on people. Rather than saying their space of knowledge expands into a global consciousness, we might say they are bombarded by images. What effect might this have? Well it might be one of de-sensitising people to the world. Philosophers like Jean Baudrillard argue that all these events change our relationship to the world outside TV. We cease to compare images to places but compare images to images – judging things in terms of their representations. If that happens then all connection to real events is lost and we would inhabit a world of simulations. This may sound extreme but in the Gulf conflict against Iraq the TV news coverage looked like a movie, while fighter pilots described their missions as being like a video arcade game. The possibility of a society of images has implications beyond television in urban design and consumer culture more generally (see Chapters 7 and 8).

Another aspect of this process is the suggestion that TV, especially fiction and adverts, tends to portray idealised lives. TV tends to offer up objects of desire – beautiful people, beautiful lives, desirable goods. Critical commentators have argued this creates impossible desires, and then offers up material and 'buyable' substitutes for happiness. These can never live up to their promises so they leave us unfulfilled. People may be affluent but they are left unsatisfied. The result of all this is to fragment collective life into isolated, individual consumers – from community to shared family viewing to the multiple-TV-owning household – each absorbed in the spectacle on their screens rather than 'real life'. At an existential level, this paints a bleak picture where, instead of Lang's alienated, industrial *Metropolis*, there are alienated post-industrial suburbs. This picture can be linked to the work of the Frankfurt school on media. They not only decried the general situation but asked who gained from it and who controlled the process. We might thus ask questions about the US dominance of media, and ask what effect it has to import US lifestyle soaps to Latin America. We might say this is

indeed not just a case of homogenisation between places but
Americanisation. We might also look at the influence of corporations
who pay for and, to an arguable extent, control a lot of media output.
These interpretations thus place an emphasis on the political economic
effects and control of media. They emphasise that broadcasting means
one voice or vision is projected to many silent onlookers with inevitable
power imbalances. Much heated debate has followed, both over media
and consumer society as a whole (see Chapter 8), between those
highlighting the potential of media for perpetuating global inequalities
and manufacturing a dreamworld that befuddles viewers and those who
suggest these accounts pay too little attention to these viewers as thinking
human beings. It is not enough to see them simply as duped by the
media.

TV as gathering place

One illustration of how we might see more active roles and geographies
for viewers would be to look at how TV functions as a 'gathering place'.
It does so on at least two levels: first, in the local communities of viewers
and, second, creating communities of viewers who may not directly
know each other. First, let us for a moment think about watching TV, say
the currently successful US series *Friends*. In it a group of under-
achieving twenty-somethings work in down-market jobs (for instance a
waitress in a themed restaurant; see also Chapter 9); they live in rather
nice apartments, and generally do rather little. The individuals are of
course quirky, humorous and all remarkably beautiful individuals. The
show is sponsored in the UK by a hair-care company, with trailers clearly
targeting twenty-something single women. But this is not the end of
watching *Friends*. On the contrary, it may be on for only twenty-five
minutes a week (once the US commercial breaks have been removed) but
it is often talked about for far longer. The conversations can in fact mock
the effort that goes into the 'hair care' or the dieting the actresses
endured to achieve their 'look'. Programmes thus form a social resource
about which people chat, gossip or argue. In this sense shows form social
events which can feed into other social occasions rather than be seen as
displacing them.

At a second level, TV programmes can also create communities out of
people who do not know each other. Some collective identities are based
around being an audience or common addressees of a message. We could

go back to radio and think of Roosevelt's 'fireside' chats to the US electorate – trying to bind people into a national consensus around the New Deal. The formal structure of the US President speaking to the nation on television also means those watching know they are thus part of the US public; the same purpose animates the British Queen's address on Christmas Day. The process is not confined to such ritualised occasions, but could instead be the network news or sporting event. The viewers know that millions of others like them are watching this occur; they are united as addressees, as a community witnessing the event. Now one might argue this is not a very strong community and that it can operate to exclude as well as bring together. This argument will be taken up in more depth in Chapter 10, but it does offer a different geography of community and belonging than the face-to-face models so often implicitly adopted.

nputer-mediated communication

One of the newest spaces created by media is the Internet – or computer-mediated communication more generally. Here many of the issues discussed over TV are raised again. We can look at the borderless flows of information heralded so often as perhaps another step in deterritorialising social life. In the case of the Internet, though, it might be argued that this is not subject to the same power relations as television. Broadcasting is an inherently one-to-many process, with a small number of producers and many recipients. The Internet offers possibilities of many-to-many interactions. An arena where people can communicate around the globe without intermediaries.

Some have heralded this as an end to geography – the collapse of distance rendering place irrelevant. Global information flows now mean that British Airways tickets for flights from Europe to America, can be bought on the phone and will be processed in India. A more scrupulous interpretation, though, would suggest space does figure in the Net in crucial ways. First, in thinking about the Net itself, commentary appeals to spatial categories and metaphors. Thus there are accounts of 'electronic frontiers', 'cyberspace' and so forth. What commentators and users describe is not conventional, three-dimensional space but a new form of space for which we need to develop new maps. Second, the possible interactions with the worlds of work and home are profoundly geographical. Chapter 3 noted the separation of spaces of work and consumption, and the segregation of space within households. The phenomenon of 'tele-cottaging', working at home and using a computer to access work materials, offers to change some of these geographies – with early studies already showing how crofters in the Shetlands can work for a firm in San Francisco, and how domestic space is reconfigured where someone works at home. Third, we might note that the net is often accused of spatial fragmentation in 'real life' just as it creates a 'virtual community'. This interpretation of what might be called 'cyberville', sees individuals sitting alone in their own houses or offices interacting across the Net as a compensation for lack of local communities – or indeed thus undermining local social institutions. Fourth, the issues raised for TV about boundary maintenance and spatial control of flows reappear over issues like pornography and state control. Fifth, there are virtual places created on the net (called MUDs, Multi-User Domains) which are often structured like houses or towns. They have virtual buildings for different activities – bars, housing, or whatever the users decide. Users construct their own places which they can decorate or furnish as they wish, and over which there are entry controls. In this virtual landscape, users can wander around, meet friends and chat and so forth. This raises a host of issues about how people create personalities on the Net, the status of interactions and whether these are any less meaningful than those conducted elsewhere. The rapid growth and changes in the Net mean we are only in the very earliest stages of geographical work on it so this list can only be an outline of issues that are beginning to be studied.

Summary

A recurring theme in this chapter has been a dialectic of media promoting group belonging and/or alienation and fragmentation. Thus we saw the film *Metropolis* responding to the social turmoil of the great cities and industrial production. Music created spaces of affect where a crowd could be together to listen or dance, and could create its own norms for that space. Equally some have argued this serves to cover over the alienation of society with a veneer of fun (see Chapter 8). Likewise we might think of the Walkman as either the most personal space or the further atomisation of the city through its fragmentation into thousands of mutually exclusive personal worlds. These tensions played through the discussions of television and computer media – and opened up issues about places and flows. From Ruttman's film *Berlin*, to CNN, to the Net media, these work through the changing relationships of what Castells (1989) called space of flows versus places. This relates at another level to issues about how mobility works as a geographical motif through a variety of music and film. The idea of music as being static (folk) or placeless (classical) can be connected to later themes in the invention of national culture (Chapter 10). The distinction of place and space, the relationship of people to territory and senses of belonging forms the subject of explicit discussion in the next chapter. This chapter has shown how media do more than simply represent a world outside: they offer different ways of apprehending it and comprehending spaces. More than this they also create mediated environments and relationships whose distinctive geographies have significant implications in today's world.

rther reading

Aitken, S. and Zonn, L. (eds) (1993) *Place, Power, Situation and Spectacle: A Geography of Film*. Rowman & Litlefield, Lanham, Maryland.

Benedikt, M. (1991) *Cyberspace: First Steps*. MIT Press, Cambridge, MA.

Burgess, J. and Gold, J. (eds) (1985) *Geography, The Media and Popular Culture*. Croom Helm, London.

Clarke, D. (ed.) (1997) *The Cinematic City*. Routledge, London.

Eyerman, R. and Lofgren, O. (1995) 'Romancing the Road: Road Movies and Images of Mobility', *Theory, Culture & Society* 12: 53–79.

Leppert, R. (1993) *The Sight of Sound: Music, Presentation and the History of the Body*. University of California Press, Berkeley.

Place of Music (1995) special issue of *Transactions of the Institute of British Geographers* 20.

Rheingold, H. (1994) *The Virtual Community: Finding Connection in a Computerized World*. Secker & Warburg, London.

Thornton, S. (1995) *Club Cultures*. Routledge, London.

7 ▸ Place or space ?

- ◉ **Sense of place and belonging**
- ◉ **Placelessness, alienation and globalisation**
- ◉ **Humanistic geography**
- ◉ **Planning, instrumental rationality and place**

Geographers make much use of the words 'space' and 'place'. School books through to research monographs are littered with these terms. But rarely below degree level are readers made to ask what these terms mean and whether they are synonymous with each other. A large debate opened up in geography towards the end of the 1970s picking up on precisely these issues, and it is the related approaches that this chapter will outline. First, there has to be a little context for the discussion, providing an idea of why the debate opened up in particular ways. The majority of this chapter will be spent exploring arguments that the modern world involves the erosion of place specificity through global forces and a consequent impoverishment of human variety and experience. This chapter will look at how some writers have attempted to distinguish the affective, or emotional, relationship people can have with places in contrast to alienation from increasingly globalised spaces.

Schools of thought

Much of the impetus for this debate in geography came from two competing versions of what geography is basically about, versions that in different forms can be traced back to at least the late nineteenth century

> As soon as we agree that the purpose of every science is accomplished when the laws which govern its phenomena are discovered, we must admit that the subject of geography is distributed among a great number of

sciences; if however, we would maintain its independence, we must prove
that there exists another object for science besides the deduction of laws
from phenomena. And it is our opinion that there *is* another object – the
thorough understanding of phenomena. Thus we find that the contest
between geographers and their adversaries is identical with the old
controversy between historical and physical methods. One party claims
that the ideal aim of science ought to be the discovery of general laws; the
other maintains that it is the investigation of phenomena themselves.
(Franz Boas, 1887, cited in Stocking 1974: 9)

The version of geography that had gained prominence in the 1960s
regarded the essence of the discipline to be *spatial science*. This version
worked on spatial models, quantitative studies and so forth, looking for
regularities and patterns in spatial phenomena that might reveal general
processes for distributing activities over space or even lead to
discovering spatial 'laws'. The contrasting vision of geography, one
championed by Carl Sauer (Chapter 2) and probably the more traditional
version, saw geography as the study of the 'uniqueness of place' or 'areal
differentiation' – that is, what makes places individual. The two
approaches have been contrasted as *nomothetic* – about predicting
regular patterns over space – and *idiographic* – describing the specifics
of places. By the end of the 1960s people were speaking of a
'quantitative revolution' having driven the nomothetic version into a
dominant position. This defined geography as the study of distribution in
space rather than particular places. In the late 1970s two trends helped
reignite this debate: one inside the discipline, one outside. First, inside
the discipline the growth of humanistic geography led to a re-evaluation
of what studying places might mean. Previously it had tended to
degenerate into regional monographs looking at the way physical,
economic, social and cultural factors (normally in that order) interacted
in a particular region. Now humanistic geography shifted the focus to ask
more incisive questions about how people related to places (see also
Chapter 4). The second trend was a critique coming from both humanism
and some forms of Marxism about the sort of thinking that dominated
nomothetic geography. Marxists such as Herbert Marcuse (1964), Max
Horkheimer and Theodor Adorno (1947) were deeply concerned that the
idea of 'social laws' and a science of society was being used as, or
unintentionally becoming, an instrument of social control and
domination. Within geography this could be seen in the rethinking by a
once prominent 'quantifier' Gunnar Olsson who in his 1975 book, *Birds
in Egg*, criticised nomothetic geography for using 'a simplified and
dehumanising conception of man [sic]' (1975: 500) and further worrying

that 'scientific methodology can be made the handmaiden of
authoritarian ideology', until 'at the end is a society of puppets with no
dreams to dream and nothing to be sorry for' (1975: 496). Humanistic
writers were often critical of Marxism but shared a concern that the
world was becoming more alienating. The calculus of rationality was
reducing diversity and restricting individuality in attempts to produce a
more orderly and efficient system. Nomothetic science was seen as
symptomatic of this process. In one of the first works to explicitly label
itself humanistic, Ley and Samuels (1979: 2–3) saw one of the tasks to
act as a corrective for scientistic approaches. These, they argued, had
divided humans into a series of quantifiable attributes and what was
needed was a more holistic, actually 'human', approach to put humpty-
dumpty together again. As Edward Relph (1981: 17) put it, such moves
expressed 'the desire to develop an alternative to what can be called the
"scientistic geography" – that is the unthinking and uncritical use of
scientific method to study all human and social matters of interest to
geographers.'

This chapter explores the connections between different ideas of
geography and the ideas of space and place. It starts by examining how
we might think about people relating to places. The first take on this will
be a look at the affective links of people and places followed by a look at
the philosophical traditions that have been used to help think about this –
phenomenology and existentialism. The following section considers
whether globalisation erodes senses of place. Taking this a little deeper,
the chapter will suggest how capitalist economies may be creating
pseudo-places. Finally the chapter poses some questions to this narrative,
first in terms of works on modernity that do not see it as some story of
declining individuality and finally by situating humanist critiques in
terms of their relationship to mass culture.

On a road to nowhere: the erosion of place

Approaches to 'place' have suggested the vital importance of a sense of
'belonging' to human beings. The basic geography of life is not
encapsulated in a series of map grid references. It extends beyond the
idea of location, and thus beyond the ken of locational science. Crucially
people do not simply locate themselves, they define themselves through a
sense of place. When asked who we are, many of us start with 'I'm a
Geordie', 'I'm a Bristolian', 'a Londoner', 'a New Yorker' or whatever.

These places are more than simply spots on the globe. The place is standing for a set of cultural characteristics; the place says something not only about where you live or come from but who you are. This can be simply a matter of stereotypes but it has more to it than that. As we go about our daily lives, we learn patterns of interaction, patterns of behaviour, that become taken for granted. We have only to move, go on holiday, undertake fieldwork to see how such patterns are very often place-specific. The continued repetition of particular sorts of behaviour comes to be associated with particular places, and newcomers are socialised into the sorts of behaviour found at those places. The result is places provide an anchor of shared experiences between people and continuity over time. Spaces become places as they become 'time-thickened'. They have a past and a future that binds people together round them.

This lived connection binds people and places together. It enables people to define themselves and to share experiences with others and form themselves into communities. One of the impetuses to study such relationships was the widespread feeling that they were in some sense under threat. And if relationships to places were undermined then so were communities and people's identities.

> We are told in countless books by architects and historians that cities have become formless malignant growths, and that the pretentious towers of glass and steel within them demonstrate all the design qualities of cardboard boxes covered in graph paper. These office blocks are apparently reoccupied daily by armies of clone-like organisation men and women, issuing from the suburban blandscape wherein lives a race of uniformly bland suburbanites, striving to indulge their materialist tendencies in the latest model of a video-recorder, a package tour to Spain or, at the very least, in the ineffable sameness of the umpteen-billionth hamburger.
>
> (Relph 1981: 13)

If it is true that specificities of place are being eroded and if, Relph continues, 'the critics are correct, our lives should somehow be diminished' (1981: 14). The paradox is that if we were to ask the suburbanites they would scarcely say their life is diminished by the affluent lifestyle described above. Relph suggests modern landscapes are ambivalent 'manifestations of technical accomplishment and widespread material prosperity' and yet at the same time of 'aesthetic confusion, ethical poverty and a disturbing degree of dependence on technical expertise' (1981: 14–15). In this sense, while modern technology creates

material affluence it endangers the affective aspects of places. The form
of planning encouraged by a narrowly 'scientistic' understanding of
places may improve living standards but produce dehumanising
landscapes (1981: 64) where the 'current pressure for efficiency and
control' and 'fast-food and suburban development which establish that
modern landscape-making with its relentless rationalism, denies feelings,
ignores ethics and minimises the responsibility of individuals for the
environments in which they live'.

Humanism, science and spirituality

Such ideas of the erosion of community and place may need to be set in
some historic context. They certainly can be traced back to the ideas of
the Romantic poets in the nineteenth century. Chapter 4 suggested how
we could see the relationship of poetry, industry and ideas of landscape
in the work of the Romantic poet Blake. It is possible to argue the
Romantic movement was a reaction to the rise of a rationalised space. At
the end of the eighteenth century, the *Enlightenment* gave rise both to
humanism in its modern form and rationalistic science, expressed in the
belief humans could shape, control and master the earth through free
inquiry and science. An example of how this impacted on ideas of space
and place was Thomas Jefferson's plans for the North American
continent. Jefferson set out a geometrical division of the then known
United States, subdividing the space according to proportions and
rational divisions, laying out the acreage of plots for townships in
carefully planned proportions, taking up the gridiron pioneered by the
Spanish in Latin America and defining what plots should go to public
buildings (schools, town halls), what to parks and what for residences.
Here, then, is a prime example of mapping *abstract space* on to territory,
dividing the land up according to an elegant principle and rational logic
hundreds of miles removed from the actual scene. Such is a *Cartesian*
vision of the world (named after the Enlightenment philosopher René
Descartes) which separates the observer from the landscape and imposes
an intellectual order upon it. 'Clearly the notion of landscape which
prevailed in the eighteenth century was closely tied to a humanism
which stressed the authority of human reason over nature. This attitude
has been preserved in scientific, technical and some academic
approaches to landscape up to the present day' (Relph 1981: 58). In
contrast, the Romantic poets looked for the sublime in the landscape,
something that would speak to them of divine beauty and purpose, of

awe of the majesty of nature. As Wordsworth put it in a 1798 poem to his sister:

> It is in the hour of feeling.
> One moment now may give us more
> Than years of toiling reason.
> (Cited in Relph 1981: 36)

This stressed spiritual experience of place rather than rational understanding. It spoke to the unique experience of place which transcended the ordinary. Such transcending moments were suggested to express the divine order. The context for the rise of Romantic ideas can be exemplified by Ruskin's sympathetic comments on the development of Romantic painting:

> Spirituality had been cast out of the world and a mechanical and materialist universe substituted; the aridity of this universe had effectively driven the Romantics to natural scenery both as a source of beauty in contrast to the ugly and false modern world of men, and for its intimations of divine order.
> (Cited in Relph 1981: 38)

chnology and the experience of space

If such was the feeling in the nineteenth century, then developments at its end and the beginning of the twentieth century strengthened the trend for abstract ideas of space to dominate. The advent of railroads meant that the sensation of travel was one of a uniform movement – separated from the world in a way coaches toiling over winter roads were not. Space could be homogenised into units of time (Schivelbusch 1977). Meanwhile railroads required more accurate time-keeping. At Bristol Temple Meads station a three-handed clock had to be installed with two minute hands, one for London and a second, eight minutes behind, for Bristol. Up to that period noon meant the time when the sun reached its zenith, which is eight minutes later in Bristol than in London. It is easy to imagine the time-tabling difficulties these local times caused railways; difficulties compounded over the vast scale of countries like the USA. So at this period we find uniform time zones imposed, and local timings begin to give way. The telegraph brought near instant transmission of information, the radio brought such transmission to thousands.

The confluence of all these trends meant that the Italian Futurists, such as Martinetti in 1908, would proclaim a manifesto dedicated to functional buildings and with a motif for a new age of speed and light (Thrift 1995). The period saw the end of ornate styles, with floral and baroque touches, and the rise of architects such as Le Corbusier, declaring the house should be seen as a 'machine for living in'. In the midst of this came the First World War, the first total war, where the technical abilities not just to create weapons but the organisational ability to support armies on a mass scale was shown to its inhuman potential. The machine age had its first machine war (Relph 1981) where people became numbers in a dehumanising slaughter, where casualties became part of an obscene balance-sheet in a war of attrition. At the same period came the rise of Taylorism, time and motion studies that broke down the workers' jobs into tiny discrete tasks, allowing jobs to be subdivided into tasks of mind-numbing repetition, and following that the Fordist assembly line heralding the mass consumer age (see Chapter 9).

It is in response to this that writers began to look on technical advance as bearing too high a price tag. The human landscapes of places to which people are attached are sacrificed to placeless, soulless new spaces which are functionally more efficient but reduce the quality of experience. Relph argues that 'further attempts to develop and apply rationalistic techniques of design and planning will result at best in incremental improvements. Indeed, with the continuing arrogation of responsibility to specialists, coupled with all sorts of remote and unanticipated consequences stemming from new technologies, they will probably do far more harm than good' (1981: 18). Scientism and the endless pursuit of technological improvement do not speak to issues of ethics or values. Indeed they have declared themselves value-free or neutral. This lack of connection by technical experts is seen as both disempowering and dangerous. W.G. Hoskins (see Chapter 2) in his most famous book, *The Making of the English Landscape* (1955), can barely bring himself to comment on anything 'made' in the twentieth century. He stops only long enough to decry the rows of mass-built Victorian houses as 'barracks' and to deplore the ugliness of the 'slug-like trails' of 'atom bombers' in the sky. The technical and rational planning of the landscape is alien to his vision of an organically growing landscape.

North American suburbs with relentless series of plots, carved out and divided up on a geometrical pattern, or the production of thousands of identical houses each sold as a dream of a home of your own' provide fertile grounds for arguing that these spaces may well destroy a sense of place just as much as tower blocks built according Corbusier's maxim. The important point humanistic geographers have made is not to blame the inhabitants, but to criticise the culture of planners – that is, a faith in scientific, technical logic to the exclusion of ideas of place values. Crucially we have already seen that inhabiting a place over time leads to its incorporation into local people's identities lending a feeling of endurance and persistence. This theme of bounding and controlling places will be explored in a later section. For now the focus may instead be on how people make spaces into 'homes' (Relph 1976: 17). The philosopher Martin Heidegger argued *dwelling* is one of the essential properties of human existence. Contrary to Cartesian ideas, humans cannot exist as free-floating minds, but rather have to exist in relationship to the world around them – what Heidegger describes as *'being-in-the-world'* (Collier 1991). The next section looks to explore these philosophical underpinnings behind these ideas of relationships to place.

enomenology/ Existentialism

One of the major philosophies drawn on in thinking about what a 'sense of place' might mean was derived from the work of Martin Heidegger and his reworking of phenomenology. Initially this was a doctrine developed by the German philosopher Edmund Husserl at the beginning of the twentieth century. The way its insights were taken up by Heidegger and later in French Existentialism provides the starting point for a theorisation that gives more purchase on experiential life. From the multiplicity of ideas associated with these philosophical theories geographers picked out three themes that seemed to speak directly about relationships to places. The first came directly from Husserl and dealt with 'intentionality', the second deals with the idea of essences, and the third comes from Heidegger and Existentialists, such as Sartre, which dealt with the situated nature of life and knowledge.

tended objects and making meaning

Husserl was intrigued by what goes to make up an object, what constitutes intelligible phenomena – hence the name of the theory. Husserl produced what is known as an *ontology* – that is, a theory of what exists. His theory saw things not simply as existing, but existing at different levels. While there was an observable world, he argued, there was more to objects than that. One of the ways he suggested this was so was through *intentionality*. For instance, take a football. It is obviously a real object but what is it? The amalgam of leather, plastic and stitching only becomes a 'football' when someone intends to kick it or use it for that game. The object only really becomes something when it is seen in light of its intended use. Thus his ontology suggests the phenomenon we call a football resides not only in the thing itself but in how we approach it. There is an *intended object* as well as a material thing. In order to work out the 'intentions' that went to form objects, Husserl suggested bracketing out preconceptions and thinking afresh about the taken-for-granted assumptions in everyday life. Such an idea opened up new possibilities in geography. Places were not just a set of accumulated data but involved human intentions as well. We should not just count how many shops there are on a high street, but what the high street means to it users. This idea that there is more to objects and things than their surface appearance – that there is a depth of meaning – has also been developed to think about the essence of things.

Essences and authenticity

The essence of a thing might be described as the characteristic that defines the object, in that sense its 'essential quality'. In terms of places this takes the idea a little further than Sauer's (1962: 321) definition of geography as dealing with facts that are 'place facts' and places that are unique combinations of these things. It suggests the idea of a depth beyond these simple facts; that there is more to place than a unique collection of things. Thinking back to Chapter 4, this has often been used to think about what is called the *'genius loci'* – the unique spirit of place. It is used to suggest that people experience something beyond the physical or sensory properties of places and can feel an attachment to a spirit of place. If the meaning of place extends beyond the visible, beyond the evident into the realms of emotion and feeling then one answer may be turning to literature or the arts as being ways people can express these meanings. Sometimes the claim has been taken further to suggest that not only do places have essences but also that one of the essential features of humanity is this relationship to meaningful places. Thus, returning to Relph's (1976: 1) work, he suggests that '[t]o be human is to live in a world that is filled with significant places: to be human is to have and know your place'. One of the questions this work raises is whether people can experience places differently or whether the essence, and meaning of a place are universal. Sometimes the phenomenological approach suggests that there is only one true, or 'authentic', relationship to a place, and other relationships are either imperfect or 'inauthentic'. The concern for authenticity comes through the ideas of Martin Heidegger. In his terminology these are ostensibly not value judgements, but generally they are heavily overlain with values, no matter how strongly denied. Thus Heidegger associates inauthentic relationships with *das man*, meaning 'the mob' or 'the crowd', a term which negatively contrasts urban life with the figure of the rustic peasant in his work (Bourdieu 1991). It is worth being careful, then, that certain political stances – valorising an idea of a rural folk – do not creep unnoted into accounts.

bedded knowledge

Heidegger can be useful since he emphasises that the human condition is not that of a rational, free-floating subject. It is very much not the subject of Descartes's *'cogito, ergo sum'* (I think therefore I am). The human subject only becomes able to think and act, argued Heidegger, through *being-in-the-world*; or, in Sartre's terms, existence comes before essence. Think of a clearing in a wood: can we say the clearing exists independently of the wood? Just so, argued Heidegger, we can't think of humans without thinking of them as embedded in the world. This has two important consequences for the argument here. First, people tend to think and act through material objects. Thus a place is a product of how we interact with it – we have different intentions towards a place if we live there, work there or are passing through on a journey. These all produce different 'places' for us. Thus Relph (1976: 5) develops from ideas of intentionality to say that consciousness is always consciousness of *something*, not free-floating, and it starts from our position in the world. This is as true of how geographers study places as about people living there. It depends how we approach a place, how we study it, what 'results' we will get. To coin a phrase, you have only to pick up a hammer to notice a lot of things that need thumping. Our knowledge about places is not independent of how we go about getting it. Heidegger thus provides one way of rethinking the different meanings places can have. Taking up this point Seamon (1980: 148) argues for a focus on the 'inescapable immersion in a geographical world' and goes on to suggest how attention must be focused on how people relate to the world at hand, so geography should 'unbury and describe this given-ness, of which people usually lose sight because of

the mundaneness and taken-for-grantedness of their everyday situation'
(1980: 149).

Perhaps more importantly is a second implication of this work.
Heidegger does not talk about intentions so much as *care*. Since we are
always engaged with the world, we must focus our attention on particular
aspects at any given time. We thus have different types and levels of care
for different things at different times. The world might then be seen as
comprising different fields of care, where distant events may be less
threatening, and distant places less essential to ourselves than those
closely involved (Relph 1976: 38). Thus our knowledge of the world is
always em-placed, it is always starting from and based around places as
centres of our 'care' about the world. This approach suggests we always
make sense of the world through the materials at hand, not from abstract
schema. Nor are objects studied independently of their contexts – rather
experience is seen as unified or holistic. Such provides a clear critique of
'scientific' studies of place. Relph (1976: 5) argues that geography is
stretched between knowledge and existence – with the continual danger
of abandoning itself to science and losing contact with its sources of
meaning. Heidegger thus locates geographical knowledge more closely
with existence than abstraction.

Geographers have used this idea of care to look at *inside* and *outside*
relationships to places. This is not simply in terms of physical
perspective, but experiential relationships and types of knowledge.
Scientific studies tend to concentrate on the outside stance, looking at a
place as an object rather than experiencing life within it. Indeed, Relph
(1976: 51) argued that '[t]his attitude of objective outsideness has a long
tradition in academic geography and is implicitly apparent in beliefs that
geography is some kind of integrating super-science or that there is a real
objective geography of places that can be described once and for all'.
The cataloguing of information about places is looking at them through
lens of 'instrumental rationality' (ibid.: 52) rather than seeing them as
organising experience. To this end Relph defines four different sorts of
space, or knowledges about space, produced by different relationships to
places. At the first step 'pragmatic' spaces are organised by our bodily
situation (left or right, up or down). Second is perceptual space,
organised through our intentions and centred on us – what we focus
upon, what we look at, thus tending to be centred on the observer.
Existential space is informed by cultural structures as much as our
perceptions – it is a space full of social meaning (see Chapter 3). This
space is defined in relationship to some human experience or task.

Finally, cognitive space is how we abstractly model spatial relationships. It would be a mistake to use only this last idea of space, as geographers too often tend to.

rritoriality and bounding places

Connected to these studies of how people have a place-centred knowledge came a concern with whether people always sought to define themselves and defend themselves (not just physically but also psychologically) through control of territory by creating a bounded (and often exclusive) domain. Taking up ideas of insider and outsider relationships to place, this work suggests people actively structure groups and define each other through creating insiders and outsiders. Peoples from around the world can be found forming groups that both control, define and are defined by territory. It is certainly not something confined to remote peoples. In the 1980s in the USA there was a trend for adolescents to assert their identity through 'tagging', marking public spaces with personal or group names. In Los Angeles, it is possible to map out whole ranges of territories controlled by different gangs – demarcated by the graffito around the city (Davis 1990).

Box 7.1

Territorial control and urban policy

The desire to control territory has formed a central part in arguments over crime and community in the city. Some commentators have looked at the current urban environment and interpreted crime and vandalism as signs of community breakdown and alienation. One of the proposed solutions for this is the reassertion of community control. This vision sees the city as a dense mosaic of interlocking communities, a patchwork of local groups policing themselves. One part of this is seen in the physical control over space. Commentators such as Alice Coleman (1985) looked at the rise of crime in cities and saw it as a breakdown of a moral order. This they identified with the rebuilding of cities and the rise of public housing projects and tower blocks. One of their arguments was that the communal land around such buildings became a no-man's-land, it was not owned by either the individual or community. One of the proposals for reducing crime was thus the introduction of 'defensible space', space to which access rights were controlled – which would give inhabitants more control over their local environment. Such devices include partitioning open spaces, controlling access and exit points to communal areas and so forth.

It is not only public housing that has become de-territorialised by nationalised planning. So too has much commercial property, in the interests of standardised products and economic rationality. Instead of a landscape such a homogeneous collection of buildings has been termed a 'flatscape' (Gurevitch, cited in Relph 1976: 57). In the terms of phenomenology such a 'flatscape' or deterritorialised landscape promotes an existential outsideness – people do not feel like they belong and thus do not care for their environment. The planning of space through abstract ideas actually militates against establishing effective communities according to this argument. To exemplify this Relph (1976: 51) cites Henry Miller:

> America is full of places. Empty places. And all these empty places are crowded. Just jammed with empty souls. All at loose ends, all seeking diversion. As though their chief object of existence was to forget.

The contrast drawn is with a sense of unique places to which people can feel they belong. The loss of bounded, controlled territory further undermines people's senses of identity, where normally people control this through relationships of 'I', 'We' and 'Other'. If 'I' is the personal sense of identity, then 'we' is the shared identity often sustained through shared relationships to places, and 'other' can be defined as outsiders (see Chapter 6). If the mediation of this process through places breaks down, then people's identities may become less stable (for a different account, see Chapter 10). The loss of a sense of belonging would make the world more alienating, since it would increase the feeling of loneliness. Tuan (1992: 36) observes there has been a steady trend of reducing the communal belonging since the Middle Ages, an increasing individuation of people producing the 'threatening awareness of being alone in a world that is ultimately unresponsive'. Unresponsive because we have fewer people bound together, linked to us by more than their interest and goodwill. Tuan (1992: 44) goes on to cite the novelist Albert Camus to explain this chilling knowledge that 'it is only our will that keeps these people attached to us (not that they wish us ill but simply because they don't care) and that the others are *always* able to be interested in something else.' Place, Tuan argues, 'helps us forget our separateness and the world's indifference. More generally speaking, culture makes this amnesia possible. Culture integrates us into the world through shared language and custom, behaviour and habits of thought' (ibid.).

obal space – eroding place?

Many of the trends towards the homogenisation of places relate to the creation of a global space through improved communications, both physical and electronic. To continue focusing on Relph's work (1976: 92), he suggests that the spread of markets bringing distant produce, the increase of highways and mass transport have undermined the idea of a locality. Instead there are all too often only moments in the spread of standardised tastes and fashions. These it is argued are not 'public' trends but 'mass' trends, not common standards developed in a locality by a community – as assumed by Sauer's idea of the cultural landscape (Chapter 2) – but developed by designers and professional taste engineers elsewhere. Ritzer (1993) argues that the McDonald's fast-food chain typifies this process. Indeed he suggests naming the process McDonaldisation of the world. The chain prides itself on producing exactly standardised products – even if in France the Bigmac becomes La Royale, the product is the same. Staff are trained to greet customers with the same phrases, the same manufactured enthusiasm and courtesy (see Chapter 9). There is a standard range of designs for the restaurant itself and a set of facades for its exterior. The characters and labels used to 'brand' a standardised product are often seen as superficial. Relph (1981) commented that the playground outside a McDonald's, 'McDonaldland', was precisely a combination of bright shiny exteriors, labelled with exciting TV characters to entice children in before leaving them disappointed at the banality of the product:

> McDonaldland epitomised everything to do with commercial strip
> development. In its brightness and its suggestion of fantasy that is not
> realised, in its superficial gloss to disguise a very ordinary product, in its
> intimations of adventure and freedom that barely obscure a precise and
> rigid organisation, and especially in its obvious and seductive appeal for
> commercial ends.
>
> (1981: 73)

The relationship to territory, and indeed to nature through the mass production of animals for fast food, is suggested to be the extreme of the technical relationship criticised at the start of this chapter. However, we can be more sophisticated in thinking through the spread on a 'placeless' society.

Meyrowitz (1985) noted the shift from cultures inhabiting specific areas to a more mobile society, so whereas people used to interact in a cultural area, relationships are now increasingly distantiated. Thus many interactions occur at meeting points, or 'liminal' border spaces between cultures. He suggests that in terms of executives and corporate elites we might study the non-local cultures of airports as these people fly from meeting to meeting. In a modern world it may be that there are fewer cultures that are 'place-bound'. Perhaps they were so closely linked to places in the past only by the limitations of communication rather than in some essential way, in which case the 'loss of place' does not really matter. Augé (1995: 34) suggests using the word place to refer to a 'culture localised in time and space' but suggests that we might see the current situation as one of 'spatial overabundance', where elements of cultures often associated with different places are coming together in the same space at the same time (see Chapter 10). It is tempting in this situation to look back with nostalgia towards some imagined past stability, but this is not a useful research tactic. Nor is examining all cultures as though they are, or should, be localised and bounded. Instead we need to look at how in some cases cultures are localised while in others there are 'non-places'. So we might look at the distinctiveness of the two, accepting that both forms exist. In this case we might look at Meyrowitz's airport lounge and say the difference between this 'non-place' and a 'place' is that the form is dominated by 'contractual solitariness', individuals or small groups only relating to wider society through limited and specific interactions, compared to 'places' where there is an 'organic sociality', where people have long term relationships, and interactions do not solely serve an immediate functional purpose (Augé 1995: 94). In these non-places our understandings are thus governed by texts, be they instructions to show your passport, time-tables, or adverts suggesting products we should buy. Such situations can be found on roads, in supermarkets, at airports, and in increasing numbers. In each case the relationship to the environment is distanced and often dominated by images – thus on motorways travel is punctuated by signposts to places the motorway now bypasses and as we look out of the car window, we are set apart and distanced from the landscape – in phenomenological terms, we are existential outsiders.

Americanisation and a geography of taste

The temptation to look for some ideal point in the past to contrast these trends to is very strong – and indeed is reinforced in the philosophy of Heidegger. However, such change also has a geography in itself. For instance much of the worry over placelessness in Europe can be seen as a worry over mass culture or the 'commodification of culture'. That is a fear that local, 'authentic' forms of culture are being displaced by mass-produced commercial forms. We can see these worries when Euro Disney near Paris was compared in the popular press to taking a blow torch to a Rembrandt – as an act of cultural violence. Mass culture has a particular symbolic geography here, where mass often means American (Brantlinger and Naremore 1991). Many of the European arguments over loss of uniqueness have to be read in the context of United States cultural and economic dominance in the latter half of this century. The European relationship to US culture industries is thus very often one of threat and loss (Morley and Robins 1993: 19).

Chapter 5 outlined how the encounter with the 'New World' helped shape ideas of what was European. But we can also find a strong romance for America in contemporary Europe. Earlier, road movies were mentioned as one way film and landscape are linked together (Chapter 6). In some cases the romance of the road movie in Europe is that it represents America as freedom from a claustrophobia felt in Europe; it offers an escape from being tied to places. Thus the film director Wim Wenders portrays America as a place and an idea, a country that privileges mobility – that invented the term and thing called a mobile home. Such an idea of mobility and being at home contrasts with the sorts of visions of belonging in a place seen above and associated with ideas of 'homeland', *Heimat* or *hembygd*. This is the characteristic Wenders finds attractive and uses so that '[t]he idea is that, not being at home [my heroes] are nevertheless at home with themselves . . . Identity means not having to have a home' (in Morley and Robins 1993: 25). Instead of thinking about the loss of community in places, we might follow ideas about the 'communion of the freeway' and look at this experience of mobility as being about a different sense of belonging (see Chapter 6). Equally the spread of the 'strip', the commercial landscape of neon and cars is not necessarily seen as a bad thing. Thus architect Robert Venturi (1973) suggested that planners needed to learn from the strip rather than despise it. In its unplanned, gaudy, commercialism running down the side of major roads it spoke to what people wanted and enjoyed. By comparison architects and planners often spoke for people, in a paternalistic manner, telling them what they *should* want. Venturi argued that the truly popular American architecture was Las Vegas, not the sort that won academic prizes. What was needed was an appreciation of the drama of the strip at night.

We must be cautious not to impose our own ideas about 'placeless' landscapes on others. Thus Campbell (1992) argues that on many urban estates with serious crime problems, the male youths often involved have an extremely strong sense of belonging, of occupying and controlling the territory. They are anything but placeless. Or in terms of landscapes that seem alienating we might do well to think of Rowles's work with the elderly. He found they had senses of place about imagined places as well – both those of memory and distant places associated with where children now lived. Thus he writes of one of his informants:

> She was reluctant to talk about physical restriction, reduced access to services, spending more time at home, problems of social abandonment, or fears for the future. Instead, as we sat in her parlor poring over treasured scrapbooks in which she kept a record of her life, she would animatedly describe trips to Florida . . . the current activities of her grand-daughter in Detroit, a thousand miles distant. She would describe incidents in her neighborhood during the early years of her residence. Blinded by preconceptions I could not comprehend at first the richness of the taken-for-granted world she was unveiling.
>
> (1978: 55)

The landscape for his informants was more than a picture or a collage – it was a 'reservoir of feeling' (ibid: 59). It is crucial then that when thinking through senses of place, these ideas are seen in a social context – about what appeals to whom, about how different people may feel they belong in different ways and value landscapes very differently.

Manufacturing difference

There is an industry that sets out to 'imagineer' places, to create 'uniqueness' in order to attract attention, visitors and, in the end, money. Landscapes can be engineered, their culture commodified for financial gain. If places are becoming increasingly alike, the rewards for standing out are increasing. Such manufactured difference often takes the form of façades placed over standardised substructures – designed to tone in with an area or to distinguish an otherwise ordinary building. This tendency, in part, gives rise to criticisms of a superficial or depthless culture – where apparently historic façades are actually modern fabrications. This would seem to contradict Relph's assertion that 'there has been a relative desacralising and desymbolising of the environment . . . particularly for everyday life' (1976: 65). Instead we might argue that increasing attention is being given to the symbolism of the built environment. These may not be the symbols of allegedly organic community or the religious symbolism of Gothic cathedrals, but symbols they are. Western cosmologies still take material form – but are now expressed through commodities (see Chapter 8). The fear of a homogeneous, rationalised world has haunted the twentieth century – the fear of what Max Weber called the 'iron cage of bureaucratic rationality'. As Chapter 6 showed, films such as Fritz Lang's *Metropolis* worried about the effects of 'instrumental rationality' giving rise to totalitarian systems, where people became simply numbers or functions. However, the increasing use of façades, the overt concentration on symbolism, suggests a different vision of society. Fritz Lang's vision hardly fits in with say the Tonga Rooms in the Fairmount Hotel in San Francisco, where the bar has been themed as a Pacific Island – complete with thatched 'hut' roofs for tables, waterfall behind the counter, artificial lake, with an island hut for the band and simulated tropical thunderstorms. To characterise this we might prefer Ernest Gellner's adaptation that this is the 'rubber cage of fictitious re-enchantment' (cited in Anderson 1990: 71). It is this shift that has led some to characterise the world as moving from modern rationalism to postmodern style.

Such simulation (discussed further in Chapter 8) can be argued to create 'pseudo-places' that only exist through the active creation of mythical spaces. Phenomenologically they are suggested to be 'inauthentic', being outside inventions rather than expressions of the culture of the locale. Their symbolism tends to be created by and directed at outsiders, so they might be called 'other-directed' (Relph 1976: 92). A prime example is Disneyland or what Relph lampoons as Vacationlands:

> The products of 'Disneyfication' are absurd, synthetic places made up of a surrealistic combination of history, myth, reality and fantasy that have little relationship with a particular geographical setting.
>
> (Ibid.: 95)

Examples of the process would obviously include such places as 'Bibleland' in the USA (Lowenthal 1984), the Parc d'Asterix in France and could also be extended to include the rise of themed pubs (say 'Irish pubs' run by a UK chain). One of the industries most associated with this trend is the tourist industry. Commentators such as MacCannell (1976, 1992) have argued that many tourist sites sell visitors an image of an 'authentic' place. That is, they 'stage' authentic places – in recreating local customs. So the Basque town of Fuentarrabia had a civic celebration of its history of independence in which all the locals took parts, but since the 1960s this has been marketed not as a festival for the locals but as one put on for the tourists (Greenwood 1977: 136). The diversity of cultures celebrated by Sauer becomes the provider of local colour to a mass tourist industry. Indeed locals may find themselves dealing with tourists to such an extent that they end up trying to look more 'authentic' than otherwise. They act so as to confirm tourists impressions of what a local should look like.

Summary

These accounts of the dilemmas over a sense of place open up many problematics that geographers are still exploring. It is a geography of modern life, or even postmodern life, that has trends of homogenisation and differentiation across the globe:

> Modern environments and experiences cut across all boundaries of geography and ethnicity, of class and nationality, of religion and ideology; in this sense modernity can be said to unite all mankind. But it is a paradoxical unity, a unity of disunity; it pours us all into a maelstrom of perpetual disintegration and renewal, of struggle and contradiction, of ambiguity and anguish.
>
> (Berman 1983: 15)

It is possible to see the challenge of current societies to 'make oneself somehow at home in the maelstrom' (Berman 1983: 345). Rather than pine for some past community, it is important to recognise that along with the loss of organic communities come new freedoms, new opportunities and excitements – the chance to escape the claustrophobia sometimes produced in closed societies, the possibilities of chance encounters and new experiences. The agenda for geography might then be to find new ways of feeling at '[h]ome in a world of expanding horizons and dissolving boundaries' (Morley and Robins 1993: 5). Chapter 10 takes these issues up by looking at spaces not as containers of cultures but as being formed from routes and from crossings of people and cultures. These will suggest that the idea of place and a single culture coinciding may be inappropriate and indeed may rely on inappropriate ideas of human experience (Augé 1995).

ırther reading

Augé, M. (1995) *Non-Places: Introduction to an Anthropology of Supermodernity*. Verso, London.

Ley, D. and Samuels, M. (1978) *Humanistic Geography: Prospects and Problems*. Croom Helm, London.

Meyrowitz, J. (1985) *No Sense of Place*. Oxford University Press, Oxford.

Relph, E. (1976) *Place and Placelessness*. Pion, London.

—— (1981) *Rational Landscape and Humanistic Geography*. Croom Helm, London.

—— (1987) *The Modern Urban Landscape*. Croom Helm, London.

Ritzer, G. (1993) *The McDonaldization of Society: An investigation into the Changing Character of Contemporary Social Life*. Pine Forge Press, Thousand Oaks.

Rowles, G. (1978) *Prisoners of Space? Exploring the Geographical Experience of Older People*. Westview, Boulder.

Sack, R. (1986) *Human Territoriality: Its Theory and History*. Cambridge University Press, Cambridge.

8 Geographies of commodities and consumption

- Spaces for selling
- Symbolic geographies and commodities
- Making worlds of goods

Until recently most geographical work on consumption was limited to accounts of retailing and distribution patterns. Recently geographers have begun to see consumption as far more than this. First, there has been a reconsideration of the spaces where goods and services are sold. Second, geographers have begun study the *symbolic cartographies* such goods and services might form. Finally, consumption is seen as including the use of goods – not merely their purchase. Put together this has meant a shift from a narrow economism, which reduced consumption to its financial bones, and a move to see consumption as extending beyond the point of purchase – see the text in this series on Economic Geography. This chapter will thus suggest consumption has its own geographies that cannot be seen as subordinate to or dependent upon those of production (see Chapter 9).

So this chapter will first look to the contexts of selling – the spaces created by society in order to sell us things. This will mean considering the traditional market, the spaces of industrialised consumption in the great cities at the turn of the nineteenth and twentieth centuries, the suburban spaces for selling in shopping malls and the turning of city centres into arenas of consumption. It will then be useful to look at the symbolic geographies of commodities themselves. This will explore how goods relate to each other, their producers and what they mean to their consumers. Finally the chapter will conclude by looking to how consumption extends onwards into the use of goods and into the home.

The specific examples are chosen tie in to themes and approaches raised in other chapters.

ces for selling

ketplace

If we are to think through the historical evolution of selling goods, a good start is the marketplace. These particular spaces have played a crucial role in the evolution of capitalistic societies. 'The market' is often used to imply impersonal and distant trade. But this misses the crucial role of the spaces through which trade operated. It was the rules and norms operating in these specific places – often interstices in feudal economies – that allowed capitalistic trade to emerge. Chapter 3 noted how the expected or required behaviour of people changed according to place and time, and the market provides an example of this. Within any market there have to be rules of reciprocity and trust – a sense of what is a fair trade. Now this is not to say everyone is honest or that reciprocal means equal, but everyone must know the score, what rights of redress they might have and so forth. These basic rules underlay the ritualised operation of the London Stock Exchange in the mid-twentieth century. There the phrase 'a man's word is his bond' expressed the idea of trust in deals made on the floor of the Stock Exchange. The market created spaces of *co-present* interaction, that is, the meeting of parties face to face. In such circumstances there are often elaborate roles and rules for interaction. In other markets rules define the possibilities of haggling, who makes what scale of offer, how long the process may go on, the turn-taking by each party to attempt to win an argument and so forth. Each party may well be performing to an implicit script. The culture of the market lies in these spatially and temporally bounded performances. This has important ramifications for thinking about geographical processes. It means 'market forces' do not operate 'out there' at some global scale, nor are they aspatial. They are embedded in local interactions and they work through these localised, spatially bounded cultures of exchange. They are not big processes which set the parameters for local interaction, rather these 'big structures' are embedded inside local interactions. Equally, 'the economic' cannot then be seen to operate as a distinct field, separate from accounts of specific cultures and social processes (see also Chapters 1 and 9).

It is worth thinking through what fairs and markets meant in a historical context and what historical markets can tell us about geographies of consumption. They were marked out spatially and temporally as places apart – market days, weekly or annual fairs. They were special occasions, to which people travelled, and by gathering created a different space than normal. A space that formed in the interstices of everyday life, outside normal rules and set apart, what has been termed *liminal*. This concept helps think through what a town fair must have been like when merchants, farmers, dealers and traders descended upon it for a few days a year. It is noteworthy to see how 'fair' is associated with markets. Thus in west Devon there is Tavistock Goosey Fair – the annual market for geese. The sale of geese has long since stopped, and it is now a fun fair. This evolution gives a clue to some of the norms operating in market-fairs. They were not just about truck and barter but were also scenes for celebration, amusement and sometimes anarchic behaviour. The fair was a place where pleasurable excesses were licensed. The spaces and times of fairs offer moments of *carnivalesque* behaviour. Such behaviour inverts the normal rules of society, celebrating excess and conspicuous consumption, a time of revelry, of gaudy display by common folk (see also Chapter 4).

Modern spaces: world fairs

The nineteenth century saw a vast expansion of capitalist markets and created new spaces for consumption. Building upon the accepted understandings of fairs we might look at the rise of 'world fairs' or 'expositions' continuing to this day as a series of 'expos'. The first of these was the Great Exhibition in the Crystal Palace in 1851. The Palace was specially built for the exhibition – a structure of steel and glass, with a barrel domed roof so it could include trees within it, letting in light from all around and yet could be disassembled and removed after the event. A special place created for a special and limited time. Although none of the goods was for sale at the exhibition, they were arrayed as a great display of commodities, celebrating the expansion of manufacturing and the scope of the capitalist market. The display was set up to enthrall and enrapture visitors with the possibilities of the economy. Planned to teach workers about the vast array of products they produced, there was little emphasis on the process of production; rather the products were displayed. The exhibition was an ideology made concrete, emphasising the achievements of capitalist production while simultaneously obscuring the conditions under which goods were produced. It thus used spectacular display to legitimise the economic order.

The title 'world fair', used for subsequent exhibitions, emphasised how products were brought from around the world, with countries having their own exhibition stands. Colonised peoples and their products were laid out as so many commodities to purchase – in a curious cross between shop and peep-show . The collection of cultures alongside goods rendered each land and people into just another commodity – at the same time their 'exotic' cultures rubbed off on to the goods – emphasising their diversity and the spectacle of the whole. New technologies such as moving dioramas could display peoples from around the world, ranges of commodities, recreated medieval towns, imagined fairylands for the entertainment of visitors. Put together such technologies of display made the whole world appear as an exhibition – a global imagination neatly captured in 'globes' in which visitors could see the world as a mosaic of goods and peoples – what Pred called a 'spectacular articulation of modernity' (1991). The coming together of industrial, mass-produced consumer goods, the power of new communications and imperial might, brought together a semblance of the whole world in one space. Such spaces then truly became a magical time and space. The magical quality is important for too often we think of selling as being about rational calculation. The appeal and power of these expositions was almost religious: a new set of rituals for a modern society, new places of 'worship', a commercial liturgy and a new priesthood of marketers leading Walter Benjamin (1974) to term these sites 'places of pilgrimage to the fetish commodity'. However, it is as well to remember the other side of fairs mentioned earlier, the ludic and the carnivalesque. From the first there was tension between the idea of an educational site, which tended to win official patronage, and the desire of audiences to have fun. Progressively more space of the fairs was devoted to 'amusements'. One effect was to blur the boundaries of consumption and leisure through new forms of visual consumption.

aces of iron and glass

However popular and large these fairs were, they were sporadic and temporary. Their influence, though, spread into arenas that were much more widespread and permanent – specifically the department store. In order to bring together an increased array of objects, department stores utilised the technologies for constructing vaulted buildings in iron and glass – to admit light and allow the circulation of people. In Paris, in the later nineteenth century, 'arcades' rose from literally covered streets, using iron and glass, to bring discrete retailers into one consumption space. There are important continuities with markets and prior spaces of consumption. A commonly accepted definition of a department store is that it brings together five or more different retail outlets or markets. The department stores drew inspiration from the use of new building techniques to produce covered markets – bringing together market stalls into a permanent enclosed space. Such markets had mushroomed in the nineteenth century, with prominent examples in London and Newcastle-upon-Tyne, leading into full-blown department stores as the owners of the buildings took control of the selling as well. We need to think through what this meant about the relationship to the goods, the practices of consumption and the meaning of the spaces through which this occurred.

These stores created 'dreamworlds' of fantastic abundance, promised the satisfaction of every need – for a price. This was not merely the case inside the shops, but also in the art of window-dressing. The possibilities of plate glass and artificial light were quickly exploited. The creation of dioramas and displays of goods became a spectacle in itself – drawing crowds to new displays. Such 'window shopping' and visual display thus impinged on the spaces of the street aiming to distract the passer-by. Many writers have argued that these displays began to change the texture of urban life. The abundant images presented, the ever-expanding display of objects of desire, affected the urban psyche by bombarding people with a vast array of visual stimuli, creating an experience of the city that was full of gorgeous fragments, full of desiring moments, but without a clear overall pattern (see also Chapter 6).

The power of these displays was regarded as such that they were blamed for a new disease of kleptomania – compulsive acquisition through stealing. This 'disease' was diagnosed as particularly prevalent in middle-class women who were the principal clientele of department stores. The increasing prominence of theft highlights how the stores were designed to create objects of desire, to simultaneously create needs and offer their solution. Emile Zola's writing (Chapter 4) suggests the ideas about gender that informed these spaces. They were designed as secure spaces, set off from the urban hurly-burly, where middle-class women could congregate in safety. They depicted these women as subject to irrational desires – hence the idea of kleptomania as a woman's disease rather than a logical outcome of the store. Meanwhile these were also highly rationalised spaces of labour, where working-class unmarried women often slept in dormitories, under strict regulations and in a highly rationalised staffing system. So these spaces brought together ideas of desire and rationality, they framed the practices of gendered socialisation and regulated contact between different classes.

These spaces shifted the terrain of public and private domains. One of the governing ideologies of Victorian urban thought (and indeed one that still lingers) was the separation of a *public sphere* (of productive work, politics, rational calculation and masculine dominance) and a subordinated *private sphere* (supposedly about domestic 'reproduction' or consumption, emotional feelings and femininity). Each of these components served to reinforce the others, to define what was appropriate feminine or masculine behaviour according to these coded spaces. The arcades and department stores turned the spaces between different sorts of retailers into an interior, making them an extension of the bourgeois house. Such served also to segregate experiences of the city by gender and class. One way of looking at this might be the idea of the *flâneur* (introduced in Chapter 4). If window shopping was one of the practices fostered through this process then visual consumption was very much at the core of *flânerie*. This practice is closely identified with the writings of Baudelaire who suggested an archetypal figure wandering for pleasure, lost in the crowd yet set apart from it, surveying the life of the city. Such an urbanite brings together many of the trends talked about above, the visual display and consumption of the city, a detachment to cope with the array of goods and the freedom to wander the city. This, it has been argued, codes the figure as masculine – gaining pleasure through watching, gazing at feminised spaces of consumption and the city.

Selling America

The same sort of analysis can be applied to a more recent consumption space – that of the shopping mall. Here we might also look for impacts in arrays of goods, the meanings of goods and forms of behaviour. A starting point might be the ideas of placelessness in Chapter 7, since artificially created shopping malls may destroy senses of place through their replication of anonymous, universal patterns (and goods) and the way they cut the consumer off from the outside world. However, many malls, from mega-malls such as West Edmonton, in Canada, through to specialist malls, have very specific place references in their design. For instance West Edmonton Mall has themed parts based on Old Orleans, or on Parisian boulevards while Stanford Mall in Palo Alto contains the following eclectic gathering of *place-images* in shops such as Crabtree & Evelyn (images of eighteenth-century life), Laura Ashley (romanticised early Victorian), Victoria's Secret (overtones of late nineteenth-century bordellos), Banana Republic (a colonial outfitter), the Disney shop (images of the 1940s). And, should this baffle the visitors, they can take a rest at a pseudo-Italian coffee bar or Max's Opera Cafe which alludes to the grandeur of imperial Vienna (Simon 1992). This suggests anything but a lack of concern with places, and rather a surfeit of spatial coordinates. But they are all manufactured and simulated spaces, which has led Shields (1989) to suggest they create a sense of *elsewhereness* – they could be anywhere, yet they strive to conjure up distant places and periods. The malls thus offer a fantasy vision to enhance the appeal of their goods, to attract the passing eye and add colour to their wares. Of course they are not the real places, nor often have they very much to do with the distant places depicted.

This effect is achieved through a landscape of carefully controlled references, one where meanings and significance are carefully drawn. In this respect they parallel landscape gardens (see Chapter 3). 'The contemporary American shopping mall is the formal garden of the late twentieth century culture, a commodified version of the great garden styles of western history which shares its fundamental characteristics' (Simon 1992: 231). If we take a similar approach and think about

---

Box 8.1

Simulating places

Jean Baudrillard (1989) refers to these place-images as *simulacra*, that is, they are imitations of things that never actually existed – copies without originals. 'Main Street USA' in Disneyland is meant to evoke a typical main street anywhere in the USA; but it is not actually from anywhere. It mobilises images people already have about what typical America was like. Indeed the effect may be to make the 'original' real main streets, the central shopping areas, seem rather a let-down, rather 'inauthentic'. Malls use the techniques of theme parks to create a sense of '*hyper-reality*' (Eco 1987) where the fake seems more real than the original. See also Chapters 6 and 7.

shopping malls as we did about gardens, we can see certain
resemblances. The Italian-sixteenth century garden was intended as a
space of delight apart from the world, just as the pleasures offered by the
mall are set apart from the city; the layout echoes seventeenth-century
French grandiose geometric design; and like eighteenth-century English
gardens they employ cultural and architectural fragments of diverse
places and eras. Instead of the garden offering the temptations of
seclusion and retirement, the mall offers the temptations of consumption.
The vistas and viewpoints discussed earlier are transformed: 'where once
the vista was of the grandeur of nature, it is now of the grandeur of
manufactured commodities, the second 'nature' of capitalist economy'
(ibid.: 241). The careful planning of the mall, its grammar of aisles and
anchor stores, the piped music and spectacle all take further the careful
cultivation of desires seen in department stores – and all leave their mark
in the iconography of the mall. These enclosed spaces set apart from the
city are perhaps the current example of Benjamin's places of pilgrimage
to the commodity; a manufactured and controlled environment, designed
not just to create wonder but to offer fulfilment through the purchase of
commodities.

The contemporary city is often seen as a place of chaos and danger –
from assault to traffic. The garden idea shows how the mall markets itself
as a haven, an enclosed space. This takes previous ideas about private
spaces a step further: malls provide their own security systems and try
and recreate the ambience of some idealised town – the mixing of people
in the streets brought into privatised spaces. (Though their much-vaunted
'sense of safety' is more doubtful than might appear. Statistics tend to be
hidden in neighbourhood figures but, for instance, the Northland Mall,
Detroit, recorded 2,083 crimes in 1985, with 1,041 serious ones
including assault, rape and robberies [Wooden 1995].) As the mall
becomes a space for general urban life, the centrality of malls to
suburban North American culture should not be underestimated; these
are far more than simply spaces where goods are purchased:

> Malls have become our contemporary town squares. Not just the
> preferred place to shop, they are also popular teenage hangouts and
> rendezvous areas for singles on the prowl. Shopping malls have become
> our fantasy cities.
>
> (Wooden 1995: 37)

For teenage 'mall rats' they are as much a site of social life as home or
school. They are the new centres of ritual and meaning rather than family
and church. Simon suggests that we need also to look at the way gardens,

like malls, acted as social centres: 'The mediaeval garden may be the most important model for the mall. Read against that garden of earthly delight and erotic dalliance, the mall is immediately a very recognizable space, the place of earthly delight sealed off from the mundane world, and for many of its visitors a place of "courtly dalliance, le jardin d'amour"' (1992: 241–2).

Commodifying spaces

Although enclosed environments have multiplied, in all parts of the city, there has also been a renaissance of the city itself as an arena of consumption. In part this is due to urban regeneration strategies that seek to cope with deindustrialisation through promoting spaces of consumption. What were once landscapes of labour become landscapes of leisure; former docks and factory sites become arts centres, are renovated for accommodation or form the sites for new festivals (Figure 8.1). In Manhattan, Zukin (1982) identified this with the return to the city by 'professionals', often in creative or media industries, taking up living in the lofts of SoHo. Conflicts can emerge over the different meanings groups ascribe to urban areas – over both residential and commercial development. Thus London's Spitalfields market redevelopment produced diverging views on whether the market should be kept as a local facility, an updated national market or as a tourist attraction. Likewise, early gentrifiers, who moved to down-market areas for cheap accommodation and a vibrant urban life, often resist the plans of developers which would price them out of the area. So in Minneapolis artists defended 'porn row' on the edge of their district since it separated them from the spiralling rents downtown. Many of the incomers seek an urban experience that owes 'more to the unmediated theatricality of medieval and early modern markets than to the calculated stage settings of the modern princes of merchant capital' (Zukin 1995: 190), an experience almost the opposite of sanitised malls.

Debate has flourished between those who place a primary emphasis on economic and capital forces in explaining these trends and those who look at the particular groups of gentrifiers. However, what is clear is that the meanings of particular spaces change over time, and what once may have been spaces for production become ones for consumption. In the case of SoHo, the lofts were originally for clothing industries, then, with the incoming artists, for small-scale production; but gentrification turned

Figure 8.1 *Brochure for Hartlepool quayside redevelopment. Place marketing and the transformation of a previously industrial area to an arena for leisure and consumption. The brochure offers to 'take you back in time to the sights, sounds and smells of an 18th century seaport'. Alongside the shops and museums is a marina redevelopment. Copyright Teesside Development Corporation.*

an area of artistic production into one of stylised consumption, where one of the things consumed was the idea of a bohemian artists' quarter as part of an unseemly rush to capitalise culture in the selling of urbane 'lifestyles' (Jackson 1995). For instance, in the 1980s in the UK, the Halifax Building Society used an advert of a twentysomething man, awaking in a converted warehouse with stripped wood floors and a balcony overlooking railways, going to a 1950s retro-style fridge to find no milk for his cats, before going out and using his card (the point of the advert) to get cash to purchase milk and papers (from a suitable urban stereotype of the cheery local paper vendor), all to the strains of 'easy like Sunday morning'. This is clearly selling a card by promoting a particular urbane culture.

As an orchestrated redevelopment strategy this can then involve acts of amnesia – erasing past associations in the landscape in order to 'market' it. An urge that is complemented by the burgeoning regeneration schemes that focus around spectacular events or waterfront redevelopments. An example is Cannery Row in Monterey. The site was the line of fish-canning factories based on Ocean View drive, described by John Steinbeck in his novels as full of 'sardine canneries of corrugated iron, honky tonks, restaurants and whore houses, and little crowded groceries', and whose inhabitants were 'whores, pimps, gamblers, and sons of bitches' or, seen another way, 'saints and angels and martyrs and holy men'. Yet cashing in on the fame of the novel, and with the demise of the canning industry, the area has been recreated, refashioned in the image of the novel for the benefit of tourists. A transformation in place and attitudes noted by Steinbeck: 'when I wrote Tortilla flat, for instance, the Monterey Chamber of Commerce issued a statement that it was a damned lie and that no such place or people existed. Later, they began running buses to the place where they thought it might be' (cited in Norkunas 1993: 58). The place is now marketed as Cannery Row, minus the whores and the workers, but with waxworks and displays to make the labour picturesque. In this way we can see the merging of the fictional and the real, the literary and lived places (Compare Chapter 4). Yet here too is a space for consumption, fictionalised till the copy seems more real than the original. Ocean View Drive was renamed Cannery Row, after its fictional version, in 1958; the actual place renamed for its fictional *alter ego*.

rtographies and commodities

This section focuses on the goods themselves rather than the spaces they are sold in. It asks questions about what we get for our money and why we buy what we do. To do this it looks at the geography of food, then 'exotic' goods and, finally, clothes. This section explores twin themes through these three examples. The first is the way commodities relate to their production and consumption across spaces; the second theme is the *discourses* and connotations that ascribe goods with meanings and the information these convey about places.

ting places

Food is surely the most fundamental consumer good, the most basic and most necessary part of life. As such it may seem far removed from arguments over 'cultural meaning'. Very often it is implied there is some sort of division between needs, that are 'natural', and desires that can be manipulated in the spaces described above. As hinted at in Chapter 1, such a division is not really tenable. If we think back to Sahlin's description of America as the land of the sacred dog – since there the dog is deemed 'inedible' – we can see that taboos and cultural shapings are extremely influential even in basic matters. It is not merely prices or sophisticated marketing that may place an arbitrary value or meaning on a good; its use is also culturally encoded – there is, for instance, nothing intrinsically masculine about trousers or feminine about skirts. The basic stuff of life is often at the centre of the strongest cultural rituals and rules – who may eat what with whom and when vary enormously around the globe.

In this section, though, we are not going to map out these dietary areas. Instead food will be an example of how consumption can connect people across spaces and, paradoxically, obscure these connections. This first theme highlights what Marx called the 'fetishism of the commodity form', by which he meant how it obscures the relationship of producer and consumer (Figure 8.2). To illustrate this let us take a tutorial from David Harvey:

> I often ask beginning geography students to consider where their last meal came from. Tracing back all the items used in the production of that meal reveals a relation of dependence upon a whole world of social relations and conditions of production . . . Yet we can in practice consume

Figure 8.2 *Advertisement for* Ethical Consumer *magazine, 1994. Copyright* Ethical Consumer *Magazine and Polyp.*

our meal without the slightest knowledge of the intricate geography of production and the myriad social relationships embedded in the system that puts it upon our table. . . . We cannot tell from looking at the commodity whether it has been produced by happy labourers working in a co-operative in Italy, grossly exploited labourers working under conditions of apartheid in South Africa, or wage labourers protected by adequate labour and wage agreements in Sweden. The grapes that sit upon the supermarket shelves are mute; we cannot see the fingerprints of exploitation upon them or tell immediately what part of the world they are from . . . We have to get behind the veil, the fetishism of the market and the commodity, in order to tell the full story of social reproduction.

(Harvey 1993: 422–3)

However, talking about a veil misses how some commodities speak loudly about where they may have come from, what places they want the consumer to think of and so forth. It can be too limiting to see goods as (a) visible surfaces and (b) invisible realities. The commodity is the overlapping of numerous different geographies forming networks of combination and disjuncture that can be seen on and of the surface. Thus newspaper feature articles stress the origins, though not the labour of production, in some foodstuffs:

Turn your kitchen into a Caribbean cookhouse and treat yourself to some tastes you've never tried before . . . Travel is the theme this spring, but if you can't get away to the fascinating places you've been reading about, you can at least cook up a little of the atmosphere in your own home, with recipes you've never dreamed of, using unusual ingredients plucked straight from the tropics.

(Cited in Cook 1996: 11)

Box 8.2

Commodity fetish: learning from the banana

A good example is the banana: in its form it shows few visible signs of which company or place it came from for consumers to decipher. Meanwhile bananas were marketed in the US for years using Carmen Miranda as figure, and indeed an icon, on labels. Her films worked as an exotic smokescreen covering US geopolitical power over 'banana republics', promoting the USA's military and political dominance in Latin America as accepted and acceptable. As such Miranda's movies helped make Latin America safe for American banana companies (Enloe 1989).

This is far from fears of a placeless world (Chapter 7); indeed it seems to offer the world on a plate (Figure 8.3). These foods are visible markers of the shifting cultures around the globe. Inspired by Harvey, I have occasionally asked a similar question of my students about a night out in Durham. Their responses often show a diverse cultural cartography – from German beer, to English ale (though that is often IPA (Imperial Pale Ale) brewed originally for export to India), infused with American rock music, with Caribbean or British influences – and the ritual finale of a curry. The only area spectacularly avoided is the local regional culture.

However, this is not to say that the global village has a happy and friendly marketplace. Cook (1996: 11) criticises the same newspaper for suggesting 'Britain's multiracial communities have brought the flavours of the world into our high streets . . . make the most of them – you need not go to India, Singapore or the West Indies – stay at home and relish them here.' It is clear from this there is an implicit 'us' who are meant to be at home and a 'them' who are meant to be foreign. Which is to deny the co-histories of black and white Britons which are often structured around these very commodities – it silences a history of Britain built around trade in commodities such as sugar, tea and tobacco that was built on black labour (see Chapter 10). Neither party's history can be grasped without seeing the others linked to it through these commodity chains. Indeed, we should be wary that the marketing of these 'place' associations does not simply suggest the world is a 'tutti-frutti cocktail of cultures' which 'erases all unpleasant stories [so that] the message becomes a refried colonial idea: if we merely hold hands and dance the mambop together, we can effectively abolish ideology, sexual and cultural politics, and class differences' (Gomez-Pena, cited in Cook 1996: 59).

Consumption and globalisation

Buying a bit of the Other

One of the geographies commodities speak to is that of an exoticised Other with the same connotations and discourses about Others found in the literature and stories in Chapter 5. One discourse in the process of colonisation was about feminised, lush tropics – where fruits of the earth grew without labour, and the indigenous inhabitants did not need to work – and the same continues to invest meanings in commodities today. In the

Figure 8.3 *J. Sainsbury's 'Taste of Mexico' advertisement, summer 1995. The text alongside the originally colour pictures reads: 'Our TexMex range has the very best in traditional southern USA and Mexican cuisine. There's hot and spicy Jamaican Jerk Chicken Thighs, the famous BBQ rib and Mesquite salmon steaks to name but a few.'*

UK there are a series of adverts for 'Bounty Bar', which generally feature a heavily eroticised beach and climax in a women eating the chocolate bar. Not only is Bounty billed as 'a taste of paradise', with its coconut filling, but effortlessly appears when a coconut falls to the ground – of its own volition – and splits open to reveal the chocolate bar. The advert in the 1990s reprises very closely the themes of feminised, sexualised, erotic places without human labour noted some three hundred years before.

These sort of geographies are not confined to food, and can be found in products as far apart as interior decor and bath products. One example is the US based mail-order and shop chain Banana Republic, which sells clothing with obvious references to colonial exploration and empire building. Lester (1992) argues that the Banana Republic mail-order catalogue casts the Other as an exotic object, using international imagery to commodify Others and thus to sell its products and a way of life, through producing a 'we' of the First World existing in relation to the Third World and thus creating an exotic other as different and distant. In particular images of colonial power in Africa are made to symbolise a relation of West and non-West, not only generalised to other places but timeless. This works to consolidate an idea of who 'we' are through excluding Others. The separation creates an imagined geography where the images foster a desire for these far-off places, a 'lack' or need that can be filled by the product. The product is meant to stand for the desired qualities. Thus if we buy a product we buy into a dream, into its associations. The Banana Republic thus positions colonised Africa and its peoples as a desirable dream, and positions it as obtainable through the commodities in the catalogue or store:

> The Exotic Other exists in a time which nominally is the present, but which is constantly represented as past: a past that is salvageable, nostalgic and mostly apocryphal. Once the Exotic Other is constructed, it is then collected, voraciously, wherever it can be found, especially in those mythic places that transform materials and peoples into commodities displayed for the pleasure of collectors and the 'We' of the collection.
>
> (Lester 1992: 79)

The narrator in the catalogue leads the reader on a roaming story where the people of the Third World feature as static cameos, and each encounter allows the reader to accumulate and collect the exotic Other – through the various commodities that symbolise each encounter.

shion and tradition

Accounts of global consumption have traditionally been bound into a series of binary pairs, where 'fashion' is contrasted to 'tradition', 'Western' to 'indigenous', and indeed mass-manufactured to handmade. However, such oppositions should not be simply accepted at face value. Traditions seem static (as in the Banana Republic catalogue above), but careful research often reveals that traditional forms have evolved continuously. Equally, what is now regarded as 'traditional' is just as likely to be inspired by contemporary trends. Many of the patterns and cuts of tartan in Scotland are a product of a Victorian Romantic revival. The traditional can provide a powerful nostalgic desire by appearing as stable and unchanging. In Bengal, Dhakari saris appeal to urban middle-class nostalgia for rural village life, employing images of 'eternal Bengali woman' and classical poetry emphasising village and home. These sort of appeals grow stronger as the world appears ever more fragmented. It is in this context that the traditional becomes first valued, so that the 'primitive' is no longer something to be surpassed but recovered; it is not 'experienced as a lack in civilization, it has been rendered available as an image-object' (Nag 1991: 106). The effect of this is that various traditional forms are resold and repackaged; consuming them creates a synchronic rather than diachronic idea of cultures in the world. That is, instead of seeing one style succeeding another over time, as in stories of progress where things steadily improve, different cultures' artefacts are available as contemporaneous choices. In this global cultural market the 'producers of culture have nowhere to turn but to the past: the imitation of dead styles, speech through all the masks and voices stored up in the imaginary museum of a now global culture' (Jameson cited Nag 1991: 105–6). We create a pastiche, a bric-a-brac of bits of different cultures and periods.

cemaking through consumption

ng goods

The connotations of products lead us on to the final way of studying geographies of consumption. Work is now opening up the constellations of meanings created through the assemblage of goods by consumers. This results from dissatisfaction with the way many analyses of advertising and commodities see consumers as dupes of 'hidden persuaders' (see also Chapter 6). How people used goods was invisible to studies that focused on point of purchase as defining consumption. Instead if we think back to the study of material culture, with which this book started, we might now suggest that mass consumption forms the dominant context through which people realise meanings in their lives and organise their relationship to the world. We have to look at the way people assemble and use commodities as well as how and where they are sold. One way into thinking about the meanings goods have in use is to look at *positional goods*. These are ones which display status within a society, thus just as much as length of penis-gourd may indicate status in Irian Jaya, so we might look at the way various brands of cars characterise different status in the West. Not surprisingly such status markers may map well on to the class structure. Indeed, to follow the sociologist Max Weber, it may be more useful to define groups by relations of consumption rather than production. If people's consumption patterns already form the way they negotiate and display their allegiances and identity, then groups showing similar patterns of consumption will probably identify with each other.

Not all goods are positional goods; many more are *information goods*, in that they may not signify status but say a lot about what sort of person the consumer is. Thus buying 'green' goods, such as phosphate-free detergent, or buying 'fair trade' coffee may indicate an awareness of the commodity chains spoken of above, but it may also indicate an awareness of how neighbours and friends will see the consumer. Such goods signal an ethical consciousness to those around the consumer as well as having effects on the supply and disposal chain. The most obvious cases are slogan T-shirts proclaiming a political cause, but virtually all goods have some informational qualities. Such information may not work through a single product but rather in the assemblage of many.

In fashioning a self-image through goods, we might see what have been termed *Diderot unities* (McCracken 1990). The philosopher Diderot was content with his wardrobe and garments, until he was given a new dressing-gown. Instantly it showed up how rather worn and faded his slippers were, so he replaced them. These then rather showed up his trousers . . . and so on, until the whole wardrobe was changed. The message is that we should not look solely at deliberate and calculated display. Rather, people assemble around them goods with which they feel at ease, and which thus unselfconsciously communicate who they are.

Thus we might find patterns of 'homologies', that is, symmetries, between taste in one field and another.

sumption milieu

The home is often the arena in which consumption takes place. By focusing on 'production', and then exchange, but rarely looking at use, most geographic theories leave the traditionally feminine domestic sphere dependent on the male-dominated economy. Seeing consumption as being about various *products* also downplays the role many of these goods have as *tools* in unpaid (mostly female) domestic labour. In each case the geographies associated with women have been seen as determined or dominated by other geographies, as secondary and dependent. We need to focus on these gendered issues if geographies of consumption are not to be sexist in their assumptions.

An example of how we might then follow this through would be studying the ideologies and constellations of goods in the domestic environment. These form dense maps of associations – photographs of holidays, furniture bought in particular stores, things inherited, things from a parental home. All these things are invested with personal meanings. Chapter 3 discussed house form; just as the Kabyle house revealed cosmologies for that culture, so too does something as taken for granted in the West as the suburban house. From its very inception it speaks about ideas of a nuclear family norm in contrast with the Victorian or Edwardian town house (see Chapter 3). The defining *positional marker* for the middle class was having at least one servant. Contrast this home, a place of lower-class labour – often single women living with the family – where the middle-class woman was defined by *not* doing household labour – with the suburban middle-class home. Suburban socio-spatial relations are marked by the transformation of middle-class women from household managers to household workers, with a decline in domestic service. Deeply bound up in these changes were consumer durable goods.

Adverts for electricity promoted the values of modernity, progress and 'scientific management'. Just as the workplace saw time–motion studies, so educational films promoted ergonomical house design and housework in disciplines through the revealingly named 'domestic science'. Consumer durables become the new positional markers for a new generation of homemakers, an importance reinforced by the emphasis the 'scientific' home placed on hygiene, with the stress on good and modern motherhood within the 'hygienic' home (a home that was helpfully defined by advertisers as using electric lighting, vacuum cleaned, with refrigerated food, which would be cooked using electricity). Imagery of the home as a castle was subtly reworked, to suggest a home under siege from germs; failures in cleanliness were signs of a failed 'womanhood'. Such an emphasis on the scientific and rational may also help explain some of the antipathies to mass consumption shown in Chapter 7. It is possible to interpret trends whereby more and more of life's needs become commodified. For instance Adorno (1993) argued one thing there is less of is 'free time'; rather, there is leisure time which is conceived through the use of various goods and services and supports a 'leisure industry'. The process has been called 'the colonisation of the lifeworld' where what was once a matter of personal relations has increasingly become mediated through commodities or professional services.

The *affluent society*, as J.K. Galbraith termed the mass consumer age, can be seen as the production of certain forms of consumption environments and needs through socio-spatial relations. At the heart of suburban ideology were geographical exclusions that segregated labour and the domestic, that created spaces that excluded different classes – a geography where women might be isolated and dependent on a male breadwinner, where increasing residential dispersal fuelled the rise of a car-owning society. Coupled with the fridge, the car allowed the rise of the weekly shop and a break from local consumption patterns. Such shifts are bound in with the time-saving devices that allowed women to enter the labour market (and required more earnings to afford) through the 'double-shift' of women's paid employment and unpaid domestic labour.

Summary

The homogenisation of mass consumption stalked the fears of social commentators who argued over the erosion of authentic places (Chapter 7). Things are, though, more complex: for places, seen as spaces made meaningful through personal histories and information in goods, are continually made and remade through consumption. Moreover, in an increasingly global world there can be increasing diversity and plurality of consumption patterns. But perhaps the paradoxes of plurality and homogenisation can best be encapsulated in an advert for the Mastercard credit card. A young man is sent out by his female partner to go shopping for a dinner party – in a series of rapid jump-cuts he is seen at a wide variety of stalls in a marketplace, each one boasting a different ethnic cuisine. As he returns with all his purchases (made with the card) including flowers for his partner, we find she has also ordered a takeaway. The narrator solemnly informs us that 'As you walk through the global village Mastercard is the universal language'. Universality, difference, repackaged markets, changing gender roles, the conquest of distance, the commodification of romance, the selling of a lifestyle. Consumption provides a lens into all this.

rther reading

Bell, D. and Valentine, G. (1997) *Consuming Geographies: We Are Where We Eat*. Routledge, London.

Bryman, A. (1995) *Disney and his Worlds*. Routledge, London.

Douglas, M. and Isherwood, B. (1978) *The World of Goods. Towards an Anthropology of Consumption*. Allen Lane, London.

Deoliver, M. (1996) 'Historical Preservation and Identity – The Alamo and the Production of a Consumer Landscape' *Antipode* 28(1).

Eco, U. (1987) *Travels in Hyper-reality*. Picador, London.

Howes, D. (1996) *Cross-cultural Consumption*. Routledge, London.

McCracken, G. (1990) *Culture and Consumption: New Approaches to the Symbolic Character of Consumer Goods and Activities*. Indiana University Press, Bloomington.

Miller, R. (1991) 'Selling Mrs. Consumer: Advertising and the Creation of Suburban Socio-spatial Relations 1910–30' *Antipode* 23(3): 263–301.

Sack, R. (1988) 'The Consumer's World: Place as Context', *Annals of the Association of American Geographers* 78(4): 642–4.

Sorkin, M. (ed.) (1992) *Variations on a Themepark: The New American City and the End of Public Space*. Hill & Wang, New York.

Zukin, S. (1991) *Landscapes of Power: From Detroit to Disney World*. Berkeley, University of California Press.

—— (1995) *The Cultures of Cities*. Blackwell, Oxford.

9 Cultures of production

- Globalisation and local culture
- Culture and roles in the workplace
- Service labour as performance
- Discipline, compliance and resistance in the workplace

Culture is not something that stands outside economic relations, or as a spin off from them. Cultures are deeply involved in the continuation of economic relations of various sorts – see the text in this series on Economic Geography. The last chapter outlined how consumption was bound up in various cultural codes which help determine the values and needs of societies. This chapter will look at cultures of production, that is, in places of work, but remember that one person's leisure may be another's labour (say, unpaid domestic labour, largely by women supporting households, or the producing of leisure goods and services). This chapter will look at industrial and manufacturing activities, the ways in which one product can be connected to different cultures of production, and finally the spaces where culture, or the provision of certain sorts of milieu, can itself be the product.

Throughout this chapter the central concern will be with cultures that operate in specific places and at specific times. Thus it is not useful to see a country as a single culture; nor a city indeed; nor in some cases, even a single firm. Rather, cultures are seen as complexes of norms, behaviours and expectations that are associated with particular places and times. In this way it is suggested particular places structure interactions, and sustain particular localised cultures. However, it will become clear that just because these cultures may be embedded in specific places and times it does not mean that they are localised in effects. Throughout the chapter there will be different relationships of local culture and global process.

The relationship between these cannot be seen as one way (the global dominating the local) and unchanging; nor indeed can 'global' processes be divided from locales through which they operate – the terms are not so simply opposed to each other.

al, community and struggle

mmunity and ways of life

One of the most obvious ways in which cultural geography has related to economic production is in the study of single-industry communities where the dynamics of employment and life are particularly and peculiarly entangled. Looking to the coal fields of the UK, we find these most 'industrial' of communities scattered amid 'rural' counties – forming a landscape of stark contrasts. The communities themselves developed a deep connection with their work. People did not just happen to work in the mine – they were miners. The job implied a whole culture and way of life – one that employed more than a million people in Britain at the turn of the century. Miners were not collected in great metropoli, but scattered in isolated communities, each focused around the pit. In such communities the strength of common bonds, through shared experiences and shared work and dependency on the mine, could build extremely strong links between people – a distinctive ethos.

We should appreciate that 'class' is not a category to march up and down the centuries but a lived relationship. Thus in studies of England's north-east coal field we find the everyday lives of people produced through their class position. Thus accounts may focus on the Co-operative store as the source of everything from groceries to mining tools. The 'netties' (outside toilets) at the back of houses meant everyone could see who was coming and going – meaning there was less sense of private space and control; life had to be more open and communal. The relative equality of workers all sharing the dangers of mining also created a sense of solidarity among the men. Meanwhile the domestic routines, of preparing baths, cleaning tiny cottages, mending clothes and cooking provided the bedrock of many women's experiences. As the novelist D.H. Lawrence noted, the hardship of the mines was widely known and the lot of the miners pitied but more sympathy was due to the women in maintaining the community. The domestic routines and an ethos of making a virtue out of necessity characterise a specific class culture in these pit villages.

Communities like these have been the strongest supporters of class solidarity – a strength in evidence during the UK miners' strike of 1984–5 where, in the face of a national security force, continued misrepresentation in the media, brutality and violence on a scale that has only become clear in subsequent court cases, and huge hardship, miners struck for virtually a year. The bitterness of the struggle and how it lasted so long in spite of the state forces ranged against the miners can only be understood in terms of the communal solidarity and way of life in the coal-field communities. To understand the industry and the politics we need to look at how these communities created unique ways of life – cultures – that supported particular forms of solidarity.

Dominance and resistance in company towns

In the first quarter of the twentieth century the southern coal fields in West Virginia were marked by incredibly bitter clashes – leading to the deployment of army troops, three declarations of martial law and armed combat between up to 20,000 miners and the hired guns of the pit owners in almost set-piece battles. The question to address is why did such a culture of resistance emerge in that place and at that time? To understand how this situation came to pass we need to look at how a particular culture evolved in the area. A rural, underpopulated area, marked by an almost peasant agriculture, was transformed in just thirty years into an industrial area linked to the world economy. The discovery of abundant coal seams coincided with the expanded demands of industry for coal power and the US navy for coal for warships. Capital from Boston, Philadelphia and as far afield as London began to pour into the area.

The miners attracted to this area initially were Eastern European migrants – who often adapted and adopted the peasant cultivation system of the area – so that even in 1924 50 per cent of miners kept cows or produce gardens (Corbin 1981: 33). The activities are not so incompatible as they first appear – particularly if we consider the detailed practices of work and their relationship to one of the crucial cultural shifts linked with the advent of the Industrial Revolution. That shift is one of time-discipline where, to orchestrate production in a factory, workers have to work at the pace prescribed by the machines and managers. In agriculture and mining the work rhythm was at this period very different. The farmer worked according to the diurnal and annual seasons, the miner was on piece-work. Underground in the cramped honeycomb of galleries, miners had to cooperate with each other – in collective activities to shore up roofs for instance – but on a piece-work rate each proceeded at their own pace. The work itself necessitated pauses in actually digging coal to move it towards the surface and prop up the cavity created by its extraction. Workers were not subject to the discipline of industrial timekeeping.

However, they were subject to oppression from the mine owners who owned the land where miners lived, the houses they lived in, the roads they used, and also paid them in 'scrip', that could only be used at the company store; they even 'owned' the state legislature that 'regulated' the industry. To enforce their authority the mine owners employed armed guards who evicted 'trouble makers' – literally throwing them onto the street. Stories abound of women in labour being thrown out of houses, of women having their breasts mutilated, men being beaten or killed. Especially targeted were any miners suspected of trying to organise a union. Most union organising in mines was based around migrants from the Welsh or English coal fields who came with expectations and knowledge about collective action. These were under-represented in West Virginia. Instead of cohesive communities, the late nineteenth century was marked by huge labour mobility – with estimates of one-third of miners moving every two years (Corbin 1981: 40).

So despite several organising drives, the mine guards beat and intimidated unionised miners until they collapsed. Notably the area was remarkably free of racial division – in tiny communities spatial segregation was impossible, piece-rates militated against unequal pay and common difficulties meant that this was almost the only area of the US mining industry *not* to have strikes *against* hiring black workers. But it was from this culture that the most violent and bloody confrontation in US labour history emerged some twenty years later. In part this was due to a social network where, although mobility remained high – indeed because of that – people had social networks and contacts that over the first twenty years of this century began to extend through- out the *whole* area. Furthermore, the efforts of mine guards ensured that when an organisation did take off it was rooted in a local and autonomous social group – who refused to heed union leaders' calls for moderation. Not least the obvious violence and injustice of the rough justice of company law focused resentment and provided visible targets. The guards, the very agents of class oppression, became the focus point for an emergence of class consciousness. Communities fought tenaciously, even though forced out of homes, and pitched against the

National Guard, due in large part to the solidarity born of their isolated and adverse situation.

Globalising firms

Working for Ford?

However, looking at the way of life in a community need not confine us to single-industry towns or isolated communities. We must think slightly differently if we wish to study other industries. This chapter suggests that the norms, practices and expectations produced around the particular locales where workforces are located are vital to understanding even the most powerful global organisation. If we look at the Ford Motor Company a set of cultures evolved with the industry. Let us start with one of the most famous moments in Ford's history – the introduction of a $5 day. This was remarkable in the early days of the car industry for it paid workers a better than average wage. But this was not philanthropy; instead Ford hoped to achieve improved production and profits. Along with the high wage went a series of expectations and demands about the workforce. For a start, the wage was only for men over twenty-one and was designed for a man to support a dependent wife; Ford reasoning that those with dependants would be less willing to endanger their earnings through industrial action. Equally a 'service department' was invented which would check up on workers, discouraging alcohol consumption and encouraging a whole series of 'good' practices: as an advert of the 1920s put it, Ford was 'Building Men, as well as Motors'. A set of norms and behaviours were connected to various consumption practices, household arrangements and so forth (see Chapter 8). Ford was seeking to create a culture of conformist behaviour with good pay and stable consumption patterns – to match his mass-production and high-productivity techniques. The one being vital to the other, forming a so-called 'Fordist' pattern that became generalised after the Second World War.

However, if the socio-spatial relations of the home were vital, so were the cultures at work. Beynon's (1973) classic study of working for Ford looked at how, by the late 1960s, there was culture of conflict between management and workers in the British plants of the Ford motor company. This was not some sudden shift from a previous idyllic relationship. There had been brutal clashes with unions and Ford had run the whole assembly line flat out to build up a stock of Model-T Fords at the end of their production run, so he could then lay off all the workers for six months (without pay) while he retooled for a new model. In the 1920s Ford was known as the 'Mussolini of Detroit' for the continual push to speed up the assembly line and the mutual suspicion among workers caused by the surveillance and spying of the 'service department'. The resulting pressure led to what Beynon (1973: 31) termed a 'Fordization of the face' – where people could appear intent upon work while actually chatting with colleagues.

Ford certainly indicated that one of the things it would prefer to avoid was a workforce well versed in factory culture – the dissimulation mentioned above – when it chose to build a plant in Liverpool, drawing in 'green labour', new to the factory system, aiming once more for men with 'family responsibilities'. Equally, recruiting officers did not seek intelligent workers; indeed, since the company did not need worker input on the routine 'deskilled' jobs, too much thought might lead to unrest. Workers and managers had a very different relationship to their jobs than in the mining communities; the workers did not see themselves as 'car workers' in the same way miners saw themselves as 'miners'. It was rather one of a range of possible jobs, and likewise the managers had to move up and out of the plants in order to get promoted. Amid the continual flow of vehicles, workers had to find ways of blanking their minds to survive, taking satisfaction in the pay packet rather than the job. So Beynon comments '[p]ressure and strain are factored into the heart of a car plant. Structured into a game where there are no real winners. In this world negotiations were often a battle, occasionally a subtle psychological war'(1973: 97) . In such a climate control over the work rate became a crucial issue – an issue that then began to give rise to the more assertive shop stewards' movement, trying to wrest some control over the conditions of work from the managers, along with continual acts of subversion, such as sloppy work, designed to lessen the pace of work on the assembly line. In these circumstances it may become clearer how a culture of conflict could evolve and come to characterise relationships in the car industry.

rking for Mazda?

However, the conventional story of the last twenty years suggests that, while such conflictual relationships dogged Western companies, Japanese firms were marked by a wholly different culture. Popular commentators have made much of the idea of cultural difference. This does not mean 'ethnic' difference, but a difference in the ethos of the workplace – one in which teamwork rather than conflict is stressed. The first Mazda plant opened in the USA and planned to create a 'third culture,' neither wholly Japanese nor American. Mazda wanted to be more productive than the US Big Three, but it went about it a very different way. In a US plant 40 to 50 seconds of every minute were used effectively, and Mazda proposed to increase this by 10 seconds a minute. If we multiply this for a plant of 2,000 workers we find it amounts to 333 extra workers (Fucini and Fucini 1990). American companies had been unable to get this much work from workers, and the relentless pressure to move the assembly line faster had led to a culture of resistance and conflict. Mazda's Japanese solution was to eliminate this culture and thus increase productivity. It would be replaced with a culture of 'teamwork' and active worker participation. This was to be achieved through groups of workers looking to see how to shave seconds off tasks themselves in *kaizen* meetings.

This led to a different recruitment strategy with a range of psychometric tests, teamwork games over rounds of interviews, and interviews conducted by workers and union staff. 'No American automaker had ever asked these questions before, the Big Three had been interested only in hiring workers to build automobiles; Mazda wanted people who could become part of a team' (Fucini and Fucini 1990: 2). They sought to avoid workers from other car plants because they feared they would have to 'unteach' them 'bad habits', and they were helped by high unemployment allowing them to select 3,500 workers out of 96,500 applicants. Instead of segregated staff functions the contracts were slim – a sign, explained one manager, that things had to be run on trust, the thick legalistic contracts elsewhere were due to managers being untrustworthy. The thin contract was meant to signal both worker flexibility and also a new relationship with management. Alongside this reduction in the number of grades went the often-noted patterns of one canteen for all – staff and managers in overalls like those on the shop floor.

The choice of workers often favoured those from service companies such as Burger King which had adopted ideas of a 'crew' who did each other's jobs and worked flexible hours – 70 per cent of recruits had a non-factory background. In the end the plant proved less different than this may make it seem. Early on workers had to negotiate the distinction of mandatory and voluntary requirements discovering that in Japan workers felt duty-bound to carry out 'voluntary' tasks. Moreover the Just-in-Time system (see Box 9.1) meant that any slacking off by workers instantly showed up – as did sloppy work. This has been characterised as 'Toyotism' (Dohse, Jurgens and Malch 1985) where the company does not need an army of inspectors. If any worker slows down the two on either side notice immediately and have to act to compensate. Thus if the parts pass from Worker A to Worker B to Worker C, and Worker B slows up, Worker A notices products piling up downstream; meanwhile Worker C runs out of parts to work on and has to react or be blamed by workers further down the line. The workers act as their own surveillance. Soon some disgruntled Mazda workers began to doubt the 'team culture' they had been told to expect:

> They were going to *kaizen* out this and *kaizen* out that, so we could be more productive. The more they talked, the more it sounded like the whole thing was just a way to squeeze more work out of every worker, with a good dose of old-fashioned paternalism thrown in to keep everybody happy.
>
> (Worker quoted in Fucini and Fucini 1990: 87)

Box 9.1

Just-in-Time (JIT)

JIT is a method of organising production designed to reduce stock inventories and aid quality assurance. It contrasts with what has been called the Fordist Just-in-Case model. In the Fordist model, each worker does one task throughout the shift as fast as possible. S/he takes the parts s/he needs from stocks and sends the finished article to stocks – from where another worker will retrieve it in due course. JIT means that parts are assembled or produced when and only when they are needed – just-in-time for the next worker to use. This eliminates tying up capital in stocks. It also means that any errors or defects show up immediately (rather than having a backlog of defective goods). Very few firms have achieved 'zero-inventory', or total JIT, but it has become a powerful model.

The number of defects began to creep up to the consternation of the managers, and workers voted for their own shop-floor union representatives rather than the ones working with the company. Perhaps we can see this culture as something of a precarious ideology, where the break-neck pace of production was enabled not by Japanese ethos, but by desperation for a $13 an hour income.

Working for Motorola?

Very often capital is portrayed as bestriding a global stage – in terms of a 'global shift' – against which local groups and even states are rendered helpless. The previous two sections suggest capital, as a series of labour relations and processes, is not culture-less; it fosters particular relationships that may come to oppose it at later dates. Equally it is also clear that it brings in, utilises and changes existing local cultures. Nowhere has this process been more stark than in the so-called New International Division of Labour. Arguably in response to labour militancy and cultures of conflict in the West, and prompted by cheap labour in areas such as South East Asia, firms relocated manufacturing plants to previously non-industrialised areas. In these circumstances it has been argued the particularities of capitalism as a cultural system, a system of values, norms and beliefs become clearer:

> Certain human realties become clearer at the periphery of the capitalist system . . . the meaning of capitalism will be subject to precapitalist meanings, and the conflict expressed in such a confrontation will be one in which man [sic] is seen as the aim production, and not production as the aim of man.
>
> (Taussig 1980: 10,11)

There are continuities with previous cases in firms seeking out 'green labour'. In the case of electronics firms in Malaysia such was encouraged by a development policy that encouraged a rural–urban drift and scheduled at least 40 per cent of factory jobs for what had been predominantly rural ethnic Malays. The multinationals setting up in the 1970s and 1980s thus formed part of a wider attempt by the government to change the ethnic and economic structure of the country – which had been marked by a legacy of British imperial policies assigning ethnic ('bumiputra') Malays to rural work, and Chinese Malaysians to work in commerce. The new policy introduced kampung, or village, bumiputra women to factory work in the newly established Export Processing Zones. By 1980 some 80,000 women were employed in these factories, of whom half were in the electronics sector (Ong 1987: 146). It was on this local culture that the production of about 40 per cent of all the silicon chips made for the US depended (Grunwald and Flamm 1985).

The choice of these women was often explained in terms of things such as their 'nimble fingers' allowing them to perform minute tasks at high speed on electronics assembly lines. This phrase implies a natural, biological affinity between these women and an industrial process devised in the twentieth century – a fortuitous piece of evolution if it were credible. Perhaps instead we should think about the culture of the kampung, where women were socialised into needlework and other crafts requiring precise but minute handiwork, and at the same time socialised into accepting the monotony of such tasks. Firms also stressed docility. However, a lack of resistance was also fostered deliberately through laws restricting the ability of unions to organise workers. Firms could simply 'close' a unionised company and open a 'new' company (producing the same product, with the same workers, in the same building) and the union would have to seek recognition all over again. The youth of workers, their inexperience and the law requiring them to leave factory work when pregnant ensured a high turnover of inexperienced workers without the self-confidence to resist managers. Equally the docility was promoted by a series of family analogies, and contact with village elders to increase moral pressure on the young women not to 'let their village down'. Interestingly this provides a further significance to the role of new cultures being created. In this case the factory was 'marketed' as an imported culture, with Prime Minister Mahatir declaring a policy of 'Looking East' to Japan for social and economic inspiration through its conformist, motivated and diligent society. 'This emphasis on cultural values rather than technological expertise presented a moral imagery to validate new labour relations and win Malay-Muslim support for a program in which Japanese companies are a major presence' (Ong 1987: 149).

Such an ideology about cultural values created a climate where the new factory conditions, where workers were staring down microscopes for eight hours to perform minute soldering, where factories ran round the clock, where large numbers of single women were outside their parental village, could be accepted. The workers could be housed in residential blocks or dormitories and issued with corporate uniform and brought to work in company buses. The preference for young workers with good eyesight – which often deteriorated quickly – produced the novel situation of unchaperoned young women with a (small) disposable income in this Muslim country. Such a situation was used, especially by American companies, to attract workers. Offers of make-up as prizes for good workers, beauty contests and so forth were used by firms in the 1970s to recruit workers. Indeed many women found pleasure in being able to sit out late at night at coffee bars in the city, to socialise as they chose. However, this produced a backlash where '[t]he most common image of the new working class Malay woman is in fact "Minah letrik" (the local equivalent of hot stuff)' or Minah karan (high voltage Minah) (Ong 1987: 146, 179). The public image plays around ideas of speed, light and heat in a series of puns to suggest a lax morality – by playing on the footloose status of single women and the language of streetwalking and prostitution, for instance. The situation was such a cause for concern public pressure forced companies to chaperone their employees more closely and reduce the promotion of Western lifestyles. But the debate and criticism has thus focused on a culture of consumption, and westernisaton, seen as a moral peril, but located outside the factories and in the lives of the women. The criticism has not been in terms of how these women are inserted in the global circuit of capital.

For the women workers, the shift in life has not just been felt around the changed social relations of consumption. Inside the factory they have been subject to a time-discipline alien to the kampung. Life is lived at the speed of the assembly line, it dictates the flow of tasks. A flow predicated on continual 'speed-ups', on Just-in-time efficiency gains, where managers used work group meetings to monitor dissatisfaction and drive workers to higher targets. These methods meshed with the cultural background where managers could draw on ideas of paternalism and parental authority to justify escalating targets – where a manager explained how parents do not merely look to be satisfied with their children but always hope for more from them (Ong 1987: 163). The contrast of self-monitored village work to this world where actions were timed by the split second, and all was conducted under close male supervision, was traumatic. Workers were subject to restrictions in toilet breaks and detailed questioning over their social lives.

In this moment of transition, the women found few recourses to the strategies developed over long years in the West. Instead they fell back on the weapons of their own culture. In the words of Scott (1984), these may be 'weapons of the weak' but they were what were to hand. The firms appealed to paternal authority, the women appealed to the parental duties of the firms. The women could use their rights to time for prayer and prayer rooms to gain some sanctioned freedom from surveillance, and just as they were subject to male supervision they could appeal to 'female problems' to embarrass their supervisors into acquiescence to requests. Women may be denied some forms of resistance by cultural expectations of obedience but equally they could draw on expectations of emotional response. Most famously they were able to appeal to ideas of emotional instability and kampung folk beliefs in responding to 'devils' in their microscopes. In the graveyard shift women who had spent hours staring down a microscope would suddenly see devils, and burst out hysterically. If the supervisors were quick they might remove the woman from the line, if not a bout of mass hysteria could envelope the whole the section and even the plant. Few sights have been more incongruous than a high-tech electronics plant of a multinational shut down until a local shaman or *bomo* could exorcise the ghosts. As Ong (1987: 201–2) suggests, these 'culturally specific forms of rebellion and resistance were directed ultimately not at "capital" but at the transgression of local boundaries governing proper human relations and moral justice', forming a moral critique that was the counterpart of hiring 'green labour' and constituting an 'indirect resistance consistent with their subordinate female status'. This was a culturally acceptable idiom of protest at that place and time, in the juncture of new cultures of individuality, new cultures of work, and new ideas of gender meeting with the old. The moment of contact illuminates what we take for granted as 'normal' economic relations as a highly specific set of cultures.

w technological spaces: linkages in the global circuit

There are three important things following from this. The first is the general point that cultures of work are embedded in other local constellations of beliefs, norms and behaviour. The second is that these situations involve cultural change but we should not look to transformation as solely involving the meeting of capitalist relations and pre-capitalist, industrial and non-industrial. Although such may show the situation in sharp relief, they do not comprise the bulk of cultures of production. Third, we should not regard these local cultures as isolated from each other. These created and productive cultures are linked geographically. To build on these points this section looks at the creation of new spaces, new cultures of work in the UK and US high-tech industry. Often these high-tech firms are in areas which could by no means be said to be non-capitalist beforehand. But equally often these firms are using the products of the labour of women in electronics factories in South East Asia. They are part of the same global system but not a uniform global culture.

Instead of the closely supervised, repetitive labour of the factory these high-tech firms are associated with 'flat' management patterns with few hierarchies and with workers often on flexi-time, working according to their own speed and inclinations. Instead of being branches in a global assembly line of a multinational, these are often small-scale firms. The workers are generally male, and claim that their work is marked by a high level of personal autonomy and is 'creative'. Such is often explicitly contrasted with the language used to describe factory production:

> The flat rural landscape around Cambridge, populated by a higher density of small, science-based companies than anywhere else in Britain, seems a long way from the desolate urban areas of the industrial north.
> (*Financial Times* 1986, cited in Massey, Quintas and Wield 1992: 94)

Looking at this 'landscape' thus tends to downplay the material links to the factories on the other side of the planet. Instead it stresses how there are few hourly waged workers, and a climate of individual achievement. Yet the actual practices of work and expected behaviour rely on long hours, way beyond those contracted, with an idea of 'commitment' that extends far over that enforced by supervisors, and a blurring of work and leisure in ways that tend to reinforce the tendency of high-tech work to be 'boys with toys'. Such may indeed then reinforce gender divides, where there is no time for the fixed commitment of childcare. The gruelling work routine is sustained in part by the 'mystique around practitioners of the new high technology' and their self-conception as an 'electronic class' (Massey, Quintas and Wield 1992: 119) who do not see themselves as fitting the models of older class relations. Such suggests a complex web of status and expectations (varying from country to country) for activities such as research. In the UK, though both research and management gain status from being seen as having left 'production' behind, the finance directors are paid more than research directors, since the latter are seen as concerned with 'theory' rather 'practice'; in France, by contrast, professional qualifications in research are more highly valued.

One could take this to the final extreme if one looked across to the USA where Microsoft has its own 'campus' with a deliberately flat management structure – so no programmer is more than three levels below Bill Gates himself. Yet it is a place that is also a hot-house, with hugely long hours being put in by staff – often for 'satisfaction', not for financial reward. This is a culture where free cola and quick meals are laid on for workers to snatch a break while they carry on working; where the firm culture is that of the obsessive, boy programmer. It is this culture that gave rise to novelist Douglas Coupland's term 'microserfs' to describe the high-tech employees of the American West coast in their carefully landscaped parks and with salaries that they don't have the time to spend. A cultural geography of work cultures would thus be careful to study the local cultures in even the most global of industries.

bour and service

obal spaces: old boys, young Turks

If many people were asked to name a 'global' force, one that showed scant regard for local places, values or indeed governments, their answer might be the global financial markets. Yet these too depend upon highly localised cultures of work. If we take the City of London we see it evolved up until the Big Bang of 1986 as a highly stratified culture – where the stock exchange resembled a club more than a global market, composed of licensed traders reliant on a culture of mutual trust – a dealer's 'word is his bond' (see Chapter 8). And the use of the masculine pronoun is not incidental: the first female stockbrokers were admitted only in 1973. It was a world of dense social networks, reliant on recruitment from an 'old boys' network of private schools and Oxbridge. This whole culture was radically overhauled in the 1980s, culminating in deregulation, when foreign firms were allowed entry to the market and old job distinctions were removed. The culture of the square mile did not so much embrace 'global forces' as import the more cut-throat culture that had evolved in the New York dealing rooms. Coupled with this went the introduction of new technologies – of dealing with information on screen, of buying and selling by phone or electronic transmission rather than face to face.

One way of looking at the resulting culture is in terms of ideas of 'masculinity'. Dealers were expected to take risks, to be aggressive and move for the kill – all activities coded as an aggressive, competitive masculinity. In such a climate it is little wonder that liberal equal-opportunities policies had little purchase – firms were looking for those who displayed what are regarded as male characteristics. A dealer who made the largest gains on the day would be termed 'the big swinging dick' – a phrase neatly evoking images of money, power and testosterone. The contrast of this to an assumed ideology of female vulnerability becomes clear when dealers react to being asked to reveal their holdings and intentions by asking, 'Do you want me to lift up my skirt and show you the lot?' Dealers would suggest that women's places were in either the bedroom or the kitchen, hiring strippers for people's birthdays, sending puerile sexist faxes to female workers and so on (McDowell 1995).

In such an environment women might have to adopt the posture of an honorary male – dressing in asexual suits, where decoration tended to be connected with a submissive femininity and earned reproving comparisons with secretarial staff – and ignoring their discomfort when in wine bars male colleagues made lecherous remarks about women passing. However, the significance of gender and personality goes deeper than this. Dealers are often engaged in work that requires them to employ their tacit knowledge, their culture on behalf of the business. Much work still relies on selling things face to face with clients and investors. For such the companies would be looking for presentable staff, where personal appearance was defined to exclude the overweight of either sex. Being in control of your own presentation and image was vital not only in competition with fellow traders but also in managing relationships with clients – an importance reflected in the body culture of gyms and fitness clubs in the city. Men might adopt a clubby bonhomie with clients while women might deliberately play a mock game of 'seduction'. What this points to is the way that workers have to adopt a series of performances in the different spaces of their work. Gay workers would adopt a heterosexual role during the day to enable them to function in the dealing rooms; all men might have to adopt a stereotypical, thrusting, macho culture. The spaces of work required people to adopt certain roles and practices in order to perform their jobs. In these cases then we might see that:

> worker's identities are not incidental to the work but are an integral part of it. Interactive jobs make use of their workers' looks, personalities, and emotions, as well as their physical and intellectual capabilities, sometimes forcing them to manipulate their identities more self-consciously than do workers in other kinds of jobs.
>
> (Leidner 1991, cited in McDowell 1995: 90)

This means that in many service jobs we have to see 'performative labour', where the thing being sold is not simply a product but also an encounter with a worker.

formative labour

The performative nature of work is particularly apparent in restaurants
and bars. Most of the studies of this have been based on *participant
observation*, where a researcher has taken part in the activities and
noted how they have had to adapt their own roles to fit in with the ethos
of the space where they must perform their duties. An early example of
such work is the study of the job of a cocktail waitress in an American
bar (Spradley and Mann 1975). Brenda Mann was employed in this bar
as a waitress. It became quickly apparent that customers, bar staff and
waitresses all had different definitions of the significance of the same
events – in effect different cultural lenses through which they saw that
space. This mapped on to the sexual division of labour that dominated
the bar – waitresses were perfectly capable of washing glasses but this
was a job done by the male bar staff. A range of tacit knowledges and
codes governed the work there. First, there was a micro-geography,
where the sexual division of tasks was mapped on to a spatial division
– with male space behind the bar and female space serving the
customers – yet the whole encoded within a general 'male space' where
male clientele dominated. Routine tasks became imbued with the
characteristics of gender so that 'the values underlying masculinity and
femininity are restated continuously each night merely by the act of
working'. The needs of the bartenders (male) assumed priority, and
tasks were organised to support them: waitresses had to learn how to
sort an order to be convenient for the bar staff, grouping types of drinks
by where they were on the bar, and then, on receiving the drinks,
rearranging them so as to give each customer the right one. Waitresses
might be called upon to help in a rush behind the bar but bartenders
preserved an almost ritual purity by never helping wait on tables, and if
the female waitress did extra work behind the bar, she would be
expected to show gratitude for the privilege. Amid the apparently
unstructured work functioned a series of social and spatial hierarchies,
between bar staff and waitresses, and between waitresses where
crossing over to serve in another's patch was a matter for etiquette and
asking permission – so as not to appear to imply that the other person
was not up to the job. Equally in relating to customers waitresses
would 'card' (that is ask for ID) women they suspected of being
under-age, but rarely do this to male customers – or risk the reproach of
the bartenders. The relationships of people in this place were thus
complex between themselves and with customers, where regular
customers had a relationship to the staff, other customers had a fond

relationship to the place, and casual drop-ins related to it as simply some place to get a drink.

These sorts of performative labour can often be involved in the themed retail spaces discussed in Chapter 8. Phil Crang (1995) studied a themed restaurant, Smokey Joe's, that offered the flavours and ambience of the Deep South of America to British customers. The restaurant sold itself both on this theme and the fast 'happening' atmosphere – an atmosphere staff had to work to create. As such staff had to perform roles playing on being happy, encouraging customers to join in the atmosphere – an 'emotional labour'. Waiters also had to mediate between customer and kitchen, customer and bar – organising their work to deliver items as fast as possible, while acting to avoid any hold-ups. Waiters might appear to make special efforts if food was delayed in order to win a higher tip, or indeed if they were abused might take revenge by spitting in desserts. The performance was thus structured round a series of presentational spaces; a *front stage* where the 'show' had to be staged as well as orders delivered and taken, and a *back stage* where waiters had to negotiate with other staff to secure their orders. More than this in the small space of the restaurant we can see the bringing together of these local interactions with the selling of an exotic culture – the Deep South – alongside the meal.

Such a confluence of cultures is shown in Zukin *et al.*'s (1992) study of New York restaurants. Restaurants form a locus of national and transnational economic and cultural flows. 'A restaurant, as a place where cultural products are created and reproduced, effects the transnational diffusion of cultural styles . . . the restaurant is a "transnational space" that processes new social identities' (1992: 106). Restaurants often form the first port of call for immigrant labour, serving local or international cuisine. The micro-geography continues where front staff – the waiters – of city restaurants are often graduates, high in 'cultural capital' or knowledge and able to play on cosmopolitan themes. The back staff are often the immigrants without the cultural capital to perform for wealthy patrons.

Those wishing to attract wealthy clients may well engage staff with the cultural capital to make such encounters work. At the mass-production end of the food industry, though, there are thousands of so-called McJobs – low-paid, low-status, routinised work producing a uniform product. No one walks into McDonald's and asks, 'What's good today?', except ironically (Leidner 1993: 45); it is not only the produce but the service

> **Box 9.2**
>
> **Cultural capital**
>
> Cultural capital is a term coined by Pierre Bourdieu. He used it to suggest that just as individuals may accumulate economic capital (marketable wealth, earnings and so on) they also possess cultural capital. It is their stock of acquired – often tacit – knowledge and skills. He suggests these are increasingly traded upon in order to gain economic wealth. There is thus a constantly changing exchange rate of different forms of cultural for economic capital and vice versa.

that is standardised. Staff interactions are carefully scripted, routinising not just the workers but the customers also (see also Chapter 7). This is true not just of fast-food chains but of more upmarket restaurants – Smokey Joe's had a list of sixteen actions to be performed by a waiter for each table they served. However, the full scripting of interactions means avoiding relying on unspoken competences. In many ways it also acts to reduce the 'emotional labour', providing a script even more than a role for the worker psychologically to shelter behind. Meanwhile the flow of products is carefully controlled by the electronic inventory linked to the tills eliminating worker discretion. The staff perform according to a 'bible' of food preparation at McDonald's almost as rigid and detailed as that used by Mazda to maximise production.

Summary

There are three reasons why these examples challenge the idea of cultures as organic communities occupying a territory. First, the tensions around issues of management and control contradict ideas of harmonious cultures that the word organic seems to imply. These are highly reflexive cultures, where people are thinking about the significance of their actions – and often thinking about them in conflicting ways. Second, the role places play in sustaining these cultures is more complex than simply as group territory. People may be adopting different roles in different spaces during the course of the day, and none of these spaces provides all that comprises human life – everyone leaves them at some point. We cannot then think 'whole ways of life' as bound up in these territories; instead spaces serve to create and reproduce expectations for specific interactions and social relations. The third reason for not relying on ideas of an 'organic totality' in thinking about cultures of production is that, as each example has shown, culture is shaped by trends in global capital. Global forces do not operate outside cultures. Rather the global forces of the modern world act through cultures embedded in specific local circumstances. Studying cultures of production suggests we recast worries about changes caused always from somewhere 'out there' to think about the links of practices 'in here' in a globalised world.

Further reading

Beynon, H. (1973) *Working for Ford*. Allen Lane, London.

Cockburn, C. (1983) *Brothers: Male Dominance and Technological Change*. Pluto, London.

Corbin, D. (1981) *Life, Work and Rebellion in the Coal Fields: The Southern West Virginia Miners 1880-1922*. University of Illinois Press, Urbana.

Coupland, D. (1995) *Microserfs*. Flamingo, London.

Fucini & Fucini (1990) *Working for the Japanese: Inside Mazda's American Auto Plant*. Free Press, Toronto.

Leidner, R. (1993) *Fast Food and Fast Talk: Service Work and the Routinization of Everyday Life*. University of California Press, Berkeley, CA.

Ong, A. (1987) *Spirits of Resistance and Capitalist Discipline: Factory Women in Malaysia*. State University of New York Press, Albany.

Spradley, J. and Mann, B. (1975) *The Cocktail Waitress: Womens' Work in a Man's World*. Wiley, New York.

Williamson, B. (1982) *Class, Culture and Community: A Biographical Study of Social Change in Mining*. Routledge, London.

10 Nations, homelands and belonging in hybrid worlds

- **Nations and cultural identity**
- **Hybrid cultural identities**
- **Cultures of contact and translation**

This book began with an exploration of the historical diffusion of cultures, their transformation over space and the landscapes they created (Chapter 2). This chapter relates the movement of cultures to the issues developed about how cultures relate to each other (in Chapter 5). Putting the two together reveals some characteristics about creating a cultural identity through a cultural area – a process which will be related to contemporary nationalism. The idea of cultural and national unity will be examined through the three prisms of imagined communities, invented traditions and cultural differentiation. The chapter will then outline an alternative where cultures are not seen as 'territorially exclusive' or homogeneous but include internal differentiation. This will stress that there is no 'essential' core to cultures but that they are always 'hybrids' formed out of interactions and movement. A first cut at demonstrating these ideas will be made through 'deconstructing' the idea of a pure, territorially exclusive culture looking at the idea of British culture. Developing this, the chapter will examine cultures of circulation, specifically looking at the interlinkages and shifts in cultures as they move around the Atlantic. Finally this will raise questions about creolisation as a process or whether we can see a plurality of cultures jostling in a marketplace (see also Chapter 8).

Blood and belonging

One of the most prominent and politically charged ways in which culture gets talked about is in terms of identity as nationality. A nationality is not simply a politico-legal status – it is also about what we believe are our social characteristics, the traits we share with fellow nationals. The mismatch between such felt bonds and the politico-legal map of the world can be seen in tensions over identity in such places as Central and East Africa, South Asia, the Balkans and Quebec. This is not the place to develop a full account of nationalism and its relationship to politics and the state; we can, however, usefully highlight how movements by Serbs, Kashmiri's and Quebecois, among others, form part of a pattern. In these movements cultural identity is seen both as a fixed object, passed from generation to generation, and as territorial where the space of the culture becomes imbued with ethnic or national ideas – forming a potent combination of 'blood and soil'. Thus territory is described in bodily metaphors, from 'fatherland' and 'motherland' or granted a personality. Very often then the cultural landscape is seen as an agent in this process – it becomes seen as the container for handing on cultural belonging. Such ethnic nationalism identifies culture with a space and the space with a people – forming a circular logic whereby one's right to belong to a space is seen as dependent on possessing the culture that is also used to identify the territory. Note how in this vision of culture and space three, sometimes contradictory, things are happening.

First, identity is defined by a spatially co-extensive culture. That is, the culture is imagined as unitary (one culture occupying a space) and bounded by that space. This is irrespective of scale so that '[w]hether 'home' is imagined as the community of Europe or of the national state or of the region, it is drenched in the longing for wholeness, unity, integrity' (Morley and Robins 1993: 6). Second, this culture is made into a thing, given a substance above and beyond the practices through which it is experienced. It is no longer the way people behave that gives rise to a label, but instead that label defines appropriate behaviour. The culture is no longer seen as the *outcome* of material and symbolic practices but instead as the *cause* of those practices – a hidden essence lying behind the surface of behaviour. Thus Theodor Adorno famously considered answering the question 'What is a German?', and noted that in fact the very question 'presupposes an autonomous collective entity – "German" – whose characteristics are then determined after the fact' (cited in Morley and Robins 1993: 6). The third move in ethnic nationalism is that this essence can be threatened, contaminated, diluted or indeed even 'destroyed' by outside forces. Seeing culture linked to identity in this way thus animates, and is animated by, a series of fears. This is then an identity of retrenchment rather than expansion – in contrast to the colonial era discussed in Chapter 5.

The nationalist impulse could be seen as part of a general human need to express control and identity spatially (see Chapters 5 and 7). However, this coalescence of culture and nationalism goes further than this general proposition. It is a specific historical process, rather than a universal need, and, although it may use the language of universal process, it works through specific political and cultural mechanisms. The next three sections investigate the specific mechanisms involved in sustaining this territorial identification. If the fears and desires differentiating a 'self' from an 'other' have been discussed in more detail in Chapter 5, then here it is time to focus on the bonds created to hold a cultural group together. The first is the linkage of people, despite distance over space in an 'imagined community', while the second looks to the dimension of time in 'invented traditions'.

gined community

The phrase 'imagined community' is taken from the work of Benedict Anderson (1983) in examining the rise of the nation and nation state. He suggests we have to see 'national' identity as a historically specific form. For instance in feudal Europe identification via lineage and fealty led to a very different texture of politics, allegiance and cultural identification – sustained under the ambit of the 'universal' (being the origin of the term, Catholic) church. People did not use a nation to frame their identity. The nation and state as a polity emerge from the sixteenth century onwards in an unsteady – and by no means inevitable – series of steps and shifts. One of the most crucial shifts was when the American and French Revolutions associated the nation, not with the body of a monarch, but with a body of people – a mass of citizens (though the United States infamously excluded the black population). It may now be so much taken for granted we have stopped to ask what such a shift entailed. It may seem natural, but it is a remarkable outcome of events that involved a lot of recasting how societies thought about themselves. Feudal systems worked through a process of vertical association, that is up and down a fairly static social hierarchy – lords followed monarchic dynasties,

---

Box 10.1

Public sphere

Public sphere is a concept often linked to an idea of a civil society. It suggests that a democratic state is composed of more than simply citizens and state, more than just state institutions. It points to fora in which ordinary people can discuss, assess and act. The public sphere is often linked to ideas of 'spaces' to which everyone has access in which people can meet as formal equals – so everyone's opinion carries equal weight – and reasonably discuss conflicts, issues and events. It should be noted that, although these polities may claim formal equality among citizens, the ability to take part in this 'public sphere' is constrained by lack of access to information (illiteracy, for instance) and it allocates the media an important role in setting the terms of any discussion. Geographers have also noted how this model involves a whole series of spaces – actual, metaphorical or both. Thus the idea of fora draws on the Roman marketplace accessible to all where citizens could meet. Other historical accounts look at cafés or other places in which people can come together. We might note that these spaces are often quite exclusive in terms of behaviour and are modelled on ones only accessible to white, straight men. Currently debates are going on over whether the Internet (Chapter 6) offers new public spaces.

---

squires followed lords and so forth. By contrast a mass society where all are 'citizens', rather than subjects, entails a horizontal identification – a sharing of identity among (formal) equals.

The model of such a sharing can be found in the idea of community. Yet if asked to think of examples of community, most people would come up with instances that worked in small-scale, face-to-face environments. How then could such an idea be stretched over space to encompass the huge numbers of people in a nation state; people we only hear about, people we only read about or learn of secondhand? Anderson (1983) highlights the importance of media, notably a mass press, allowing news of events and people to spread – and to spread intact rather than through a process of Chinese whispers through intermediate levels. So people learn about the *same* events in the *same* manner. They do not have to rely on intermediaries to act on their behalf; thanks to the press, information is disseminated to the whole 'public sphere'. But this does not solve the problem for creating a sense of 'horizontal' public belonging entirely. Everyone may now relate to the same stories and arguments, the same heroes and villains, but how do we relate to our co-nationals – the people whom we will not only never meet but those whom we will never see or hear? This is where the 'imagined' part of the community comes in.

Think of reading the newspaper (or for that matter watching the national evening news; see also Chapter 6): what is important is not just that large numbers of people are reading or hearing about the same things over a large spatial scale but also that each person knows (or at least believes) others are doing likewise. It is in this belief that the shared dimension of an 'imagined community' works – it is sustained by the belief that it exists. In this sense it can be compared to the ideas of an 'intended object' (all Chapter 7).

This kind of analysis can be applied more widely than the daily press. Nation states are associated with the creation of a range of institutions – the media, school education and a host of others – which address each citizen uniformly and also convey the impression that every other citizen is also being so addressed. It is thus the uniformity of address which is crucial. Thus in the 1950s all French primary school children would use the set text of a 'tour of France' where they followed the exploits of two young characters as they travelled round the country; the education minister could look up from his desk at a particular time and day, and say 'Our children are currently crossing the Pyrenees'. The analysis of imagined communities tells us that the ability to thus identify 'our' children by their shared exposure to this education is as crucial as anything they might learn about Pyrenean France.

enting tradition

However, we should not ignore the content of national cultures. National identity often relies upon a shared history as ground for both commonality and defining characteristics of the people. The shared history brings together the above mass relationship – the 'demos' of a witnessing public – with the 'ethnos' of particular cultural identity. Few more obvious examples appear than the British polity which extensively utilises the trappings of history, recalling past struggles and achievements. However, a scrupulous viewer would notice that many of the rituals are rather recent inventions. Thus we might look to the Royal Family providing not historical continuity but uniting the British nation(s) through being a common object of discussion. The public Royal weddings of the post-war era are very recent ideas. Royal weddings up until this century were often private affairs – aimed at linking ruling dynasties across Europe, rather than at binding together the populace. The transformation into a spectacle works at drawing people together –

not only watching the same event but knowing others are doing likewise. In this way the importance of commentaries on 'how many millions' are watching, or the power surge during a tea-break (everyone is having a cup of tea – rolling one national symbol into another) become apparent. What is created is a watching community. To this extent it does not actually matter whether members of a Royal Family wish to get married, divorced, invested in this or that order – what matters is the way they provide a shared experience for a nation. The collective outpouring of grief and displays of mass emotion around the funeral of Princess Diana provide an object lesson in this idea that it is in the audience watching events that shared identity is created.

The idea of invented traditions goes somewhat further than this. It points out that while the above is true, the rituals form a certain idiom in how they portray the nation. Thus take the investiture of the Prince of Wales, performed with fanfare at Caernavon Castle. Tributes were paid to the Prince in the form of feudal gifts and the Prince's pledge to the sovereign under an outdoor canopy before assembled dignitaries (and the watching national press). All this appeals to feelings of antiquity – stressing the role of the monarchy as a symbol not just for people to gather round today, but of continuity with the past generations. Except, the whole ritual was made for television – an invented hotchpotch of forgotten events, legally meaningless heraldry-babble (written specially for the occasion) and the whole directed by a photographer who was then married to a royal princess. The striking element is thus that it is not any actual relationship to a shared history that is important, but the idea of *pastness*; the symbols of antiquity appear far more important than actual continuity. The quest for authentic national cultural identity often results in efforts to reconstruct a lost national ethos as though it were some secret inheritance or that cultural identity were a matter of recovering some forgotten or 'hidden music'. Although tradition appears as a coherent body of practice and customs handed down over generations, it is often retrospectively invented. These invented traditions reinforce the idea that national identity can be passed down over generations as though it were some precious essence and that the rituals are a container for a pre-given national identity.

Chapter 6, mentioned how folk music was often rediscovered in this way. The same can be extended to much interest in folk culture – both artistic and material. An example might be the rediscovery of folk culture in Sweden in the early years of this century. The context is significant. Sweden from a peripheral position in Europe was experiencing rapid

industrialisation and urbanisation; vast emigration to the USA marked
Swedish society; Norway had seceded at the start of the century; and,
finally, improved communications were opening up isolated rural areas.
The result was a boom in interest in preserving folk culture, resulting in
the Skansen open air museum in Stockholm and hundreds of local
institutions, each depicting their regional culture and committed to
preserving the sort of cultural landscapes the Berkeley school studied
(see Chapter 2). Meanwhile, two artists (Karl Larsson and Gustav
Ankarcrona) designed a female 'national dress' of blue with yellow
embellishments based·on folk types. The rediscovery of such roots was
thus bound up with how Sweden found itself emerging as a modern
nation state. The past was refashioned to meet the insecurities and needs
of the present; in a time of urban transition and rapid change the appeal
of 'unchanging' tradition was considerable. We might draw connections
with how Britain is coping with the post-imperial, post-industrial world
by turning to industrial heritage. Thus the boom in preserved
warehouses, industrial museums, conserved waterfronts and so forth has
been linked to an uncertainty about the future and about what it means to
be British. It has been suggested that lauding these past achievements
helps bolster a sense of security in the face of an uncertain future.
Rediscovering heritage as a way of reaffirming present identities,
particularly in times of rapid change or uncertainty, seems widespread. It
seems to function as a backward-facing mirror that presents people with
the image of themselves in the secure and stable identities they want to
see. Whether this is healthy or a true depiction of past is discussed later
in the chapter.

ltural differentiation

If the appearance of continuity is one way in which the idea of coherent
ethnic or national culture is supported then another can be found in the
differentiation of that culture from others around it. Chapter 5 explored
this idea in colonial ideology and the relationship of the West to other
parts of the globe. In this section it is important briefly to recap on the
principal ideas behind this and discuss how they more generally affect
the creation of national identity. The previous section outlined how the
idea of inherited traditions can sometimes come to be seen as a 'secret
music', audible only to that culture – defining membership – but this
does not say how such then relates to other cultures, with their own
implied hidden music. The process suggested in Chapter 5 is one of

projective identification or 'Othering'. In this process other cultures are used as a mirror to reflect our own traits, not in a happenstance fashion but as one of the mechanisms that defines those traits. Cultures contain a vast range of practices of which only a tiny minority become picked out as crucial to identity. Ideas of 'Othering' suggest these are chosen for the way they can be used to provide differentiation from others – they form a constitutive outside or a defining boundary. Cultures are not thus defined solely by what is internal to them but by how they constitute themselves over and against other cultures.

One metaphor for this process comes from the work of the psychoanalyst Sigmund Freud, later elaborated by Jacques Lacan. In studying small infants they developed a model of how children form an identity for themselves between six and eighteen months of age. Up to this point they found that infants had no clearly formed idea of the self; they were a bundle of desires, needs and feelings without being orchestrated into a coherent whole. But then a crucial development occurred in what they called the *mirror stage*. Imagine a child suddenly looked into a mirror and recognised itself – suddenly it would see a whole body, a person. Lacan argued that society worked like a mirror – reflecting back an image of an individual. In this way our sense of who we are is not based on a wholly internal process but relies on an external reflection. Much the same argument could be made about cultures – that it is by looking at a 'mirroring' other they define what it means to be themselves. However, this is not simply a neutral process in deciding characteristics. Thus everyone wants to think the best of themselves and their own cultures, but, by and large, everyone is a mix of better and worse characteristics. *Projective identification* is a term for the way in which we tend to displace – or project – the worst sides of our own identities on to others, to make them the bearers of, or responsible for, our failings. Thus we might look at the history of the West as an 'Enlightened culture', organised around self-declared goals of democracy, progress, knowledge and rationality – and how it depicted Africa as dark in contrast to light, as ignorant, as irrational, or Asia as despotic and unable to develop or progress (see Chapter 5). These powerful binary definitions work by valuing one side (the West) as embodying virtues and consigning every opposing vice to the rest. In formal logic, the structure is one of *a* against what is *not-a*. Such should highlight that none of these are neutral characteristics simply belonging to the cultures, rather they are framed and organised into these relations.

A somewhat different history could be told if the West was identified with war, invasion, forcible occupation, rule by the sword over colonised populations and so on. This latter vision is neatly encapsulated in the comment by Mahatma Gandhi. When a reporter asked him what he thought of Western civilisation, he replied that he thought 'it would be a jolly good idea'. This little anecdote raises the issue of how we identify our culture as containing certain characteristics and how such ideas of internal virtues are often tied to a history of Others outside, against whom these ideas have been framed. The idea of culture as container places a premium on 'purity' within cultures, as an essence that is handed down, and on boundaries between cultures maintaining this. However, there are often material and symbolic connections that mean we cannot simply look at cultures as bounded entities, nor can we simply focus on the differences between these entities. Rather we need to look at how such differences are constructed and often conceal not just similarities but material and symbolic connections.

tures of connection and contact

side histories

The previous sections outlined how national cultures are often thought of as containers with, first, a contents that can be passed down, that is the sole preserve of one nation, and, second, spatial differentiation into discrete, pure cultural areas – what Paul Gilroy (1993: 7) calls 'the tragic popularity of ideas about the integrity and purity of cultures [and] the relationship between nationality and ethnicity'. This section will outline how such an idea often hides the outside history of cultures, the 'outside' without which what is on the inside would make no sense. If cultures are relational then we need to explore how those relations become hidden or repressed when cultures are described as though they were homogeneous and bounded. This view of cultures leaves us looking at contemporary debates over immigration and a multi-ethnic Europe through a particularly strange idea that the meeting of black and white cultures is

> a collision between fully formed and mutually exclusive cultural communities. This has become the dominant view, where black history and culture are perceived, like black settlers themselves, as an illegitimate intrusion into a vision of authentic British national life that, prior to their arrival, was as stable and as peaceful as it was ethnically undifferentiated.
>
> (Gilroy 1993: 11)

Such an idea serves to downplay the cultural heterogeneity in countries such as Britain, through a contrast with an 'alien' culture. But, as Gilroy suggests, the relations between these cultures are far more lengthy and intimate than this portrait allows. Edward Said (1992) reading Jane Austen remarks on how the landed gentry who populate the novels are deeply linked to the Caribbean as absentee owners of (slave) plantations. The country house comes to stand for a homogeneous English culture (note the downplaying of Celtic identity; see also Chapter 3), symbolising a people marked by the Mother of Parliaments, democracy and liberty before a free judiciary – a polity defined through the public sphere. Schopenhauer asked how Britain lived up to its own cultural ideals if we were to judge it on its promotion of Negro slavery, the ultimate object of which was sugar and coffee. British self-conceptions worked by repressing the 'outside' relations of colonial power. Thus Turner on the one hand painted pictures commissioned by Caribbean plantation owners, to celebrate a vision of the rural as the distillate of national life. On the other hand he painted slave ships throwing overboard the dead and dying as a storm approached. The art critic Ruskin was only able to bring himself to look at this latter as a study in aesthetics of water painting (Gilroy 1993: 16). The way the power relations disappear into aesthetic studies neatly encapsulates the repression of linkages between cultures; Ruskin could not face the micro-cultures and transitional spaces of the ships linking these two cultures into a whole.

These sorts of approaches suggest we cannot see cultures as discrete containers, but must recognise their mutual entanglement. Thus in white culture, we can recognise how the category of 'British' gained prominence in an imperial context – colonists might be Scottish or Welsh at home but were British abroad. Equally it is too easy to forget that Britain claimed its empire as part of itself. The colonies were not separate but part of British economic and political space. The history of Britain is bound up with them, as is theirs with Britain. One cannot understand the one without the other. So Stuart Hall discusses his symbolic belonging and ethnicity:

> People like me who came to England in the 1950s have been there for centuries; symbolically, we have been there for centuries. I was coming home. I am the sugar at the bottom of the English cup of tea. I am the sweet tooth, the sugar plantations that rotted generations of English children's teeth. There are thousands of others beside me that are you know the cup of tea itself. Because they don't grow it in Lancashire you

know. Not a single tea plantation exists within the United Kingdom. This
is the symbolisation of English identity – I mean what does anybody
know about an English person except that they can't get through the day
without a cup of tea? Where does it come from? Ceylon – Sri Lanka,
India. That is the outside history that is the inside history of the English.
There is no English history without that history.

(Hall 1991: 48–9)

Making these connections between histories that are normally kept apart
are thus vital to understanding the social and cultural developments in
context. That context is one of contact and the circulation of ideas and
people through the symbolic and material networks of empire. It is
possible to trace the movement of ideas of working-class radicalism back
and forth across the Atlantic – born in the same ships with the goods and
produce of trade. It should not be surprising then that black radicals
preached in London at the end of the eighteenth century – above all since
a quarter of the British navy was made up of black sailors (Gilroy 1993:
12). The linkages of cultures, often seen as discrete containers of some
cultural essence, are profound. Studying the connections breaks down the
idea of inside and outside and opens what we could call a *third-space*
(Bhabha 1994). Not outside or inside but connected to both, what Gilroy
calls the 'double consciousness' of occupying a space between 'two great
cultural assemblages'. The interesting question then moves from simple
differentiation, from ideas of cultural areas to 'cultures of circulation',
cultures in contact with each other, always on the move and mutating.

ltures of circulation

The first step in rethinking geographies of national and ethnic cultures
might then involve shifting how we tend to classify ideas and practices.
Traditionally classification has worked on lines of distinction, splitting up
ideas and a hierarchical classification – sub-categories within other
higher-order categories, one explanation or behaviour dominant over
another. Much of the logic of classification has in fact been to create
externally distinct and internally homogeneous categories. This is a
symmetrical process to that criticised for cultures in the previous section.
So it may be necessary to think differently in order to produce different
interpretations. One set of ideas that has been suggested as offering a
way ahead here are those of the French philosopher Gilles Deleuze. He
suggested that traditional classification is bound up in an 'arborescent'
metaphor – that is, in branching tree diagrams of hierarchical, mutually

distinct categories – and thus looks to identity as a system of *roots*. In contrast he suggests thinking of *routes* of identity formed through connections and traverses, as mobile and changing all the time rather than static, as mixing and bringing together rather than dividing up and splitting categories apart (Deleuze and Guattari 1987). This sort of logic might be termed *rhizomatic*, after the growth of, say, brambles which send out shoots to produce a tangle of plants each criss-crossing the other.

How does this help us in thinking about cultural geographies? Well, let us take the ideas of tradition that we saw mobilised the past to provide a buried essence or 'hidden music' only accessible to those on the inside of the culture – it was an internal history belonging to the one group. But if we look at the forms of music created through the transportation and translation of cultures across the Atlantic (see also Chapter 6) we find a continual process of mixing, adapting and cross-fertilisation. The recovery of this spatial history of contemporary forms involves more than just looking for roots; finding the routes through which forms have propagated reveals the connectedness and mutual imbrication of different cultures. Gilroy argues (1993: 75) the emphasis on music and performance in black culture was a direct result of oppression during slavery which punished literate forms, leaving music as the only way to confront 'an unspeakable but not inexpressable barbarity'. From thence we can find the continued translation and circulation of this music – from spirituals into blues, following the mass migrations north to cities such as Chicago – songs of hope and longing as well as consolation. Without this there would have been no rock and roll, which took rhythm and blues across the Atlantic to Britain where bands such as the Beatles or Rolling Stones reworked it again. Parallel journeys could be traced in the spread and change of jazz music, and the relation of all this to the paths already blazed by spiritual choirs who travelled the Atlantic at the end of the nineteenth century. Whole geographies can be written around specific genres, indeed specific composers. In Chapter 2, a brief example was given of the changes in Acadian material culture as settlers crossed the Atlantic. We might also look at the removal of the settlers to the south due to persecution and their development of Cajun music – a music developing through their successive movements and one now seen as under threat from more freeform styles. Gilroy (1993: 95) offers a set of linkages round a Chicago male harmony band – the Impressions. The group spawned imitators round the Caribbean including the Wailers – who went on to develop avenues into Ska and Reggae, meanwhile the old soul songs of the Impressions have been taken up and covered anew by

Brummie toaster Macca B and singer Kofi in 1990. The Caribbean development of sound system music and soul, often with rap and hip-hop from the cities of the West and East coast of the US – has been taken up by South Asians in the UK – fusing Punjabi and ragamuffin.

This is no story of an unchanging essence or some hidden key but '[t]he very least which this music and its history can offer us today is an analogy for comprehending the lines of affiliation and association which take the idea of the diaspora beyond its symbolic status as the fragmentary opposite of some imputed racial essence' (Gilroy 1993: 93). This is not a simple celebration of diversity; instead it allows us to look at the particular junctures and conjunctures that give rise to particular meanings and forms. Thus rap music is marked by strongly sexualised and often misogynist lyrics – say, where women are referred to as 'bitches' on a routine basis. The connection of race and gender should not surprise us, given the connections of erotic desires and fears outlined in Chapter 5; nor should the way the media pick this up be surprising for the same reasons. Gilroy suggests that what drives this formation is the constellation of gender, masculinity, subordination and race that means '[a]n amplified and exaggerated masculinity has become the boastful centrepiece of a culture of compensation that self-consciously salves the misery of the disempowered and subordinated' (1993: 85).

ties in the world

We can see cities as places where these paths come together, cross, mutate and develop. The bringing together of different traditions results in hybrid forms. Such is not simply a bland relativism, where we say anything goes. We might instead think of the term as *creolisation* – from colonial societies where elaborate systems emerged to deal with inter-racial forms. Such perhaps suggests the politically charged nature of process, as well as evoking its often unequal starting points. Instead, though, of seeing this as a loss of purity, it can be seen as a productive situation. In this way this section has looked to Gilroy's work on music to provide a new way of mapping cultures so that the

> [c]ritical space/time cartography of the diaspora needs therefore to be readjusted so that the dynamics of dispersal and local autonomy can be shown alongside the unforeseen detours and circuits which mark the new journeys and new arrivals that, in turn, release new political and cultural possibilities.
>
> (1993: 86)

Such an approach originates with thinking through the predicaments and possibilities of being a diaspora – that is, with being displaced, permanently out of place – yet in the modern world we might instead think of such a condition as being pervasive of most peoples. Cities stand at the junction of so many cultures that we cannot any longer counterpose the local or the authentic to the global – as though the latter were some homogenising force. Instead there is the continual work of transculturation. In the music of Dick Lee, in Singapore, we find such connections – an artist educated as a fashion designer in London, proud of 'Singlish' as a fusing of languages, writing songs called Modern Asia celebrating a sense of Asianness that draws on the concrete experiences of life in Asia without seeking some timeless or lost past of mythical samurai and geishas (Kong 1996: 285). This is not losing a sense of local specificity but reworking it, so in his song 'Life in the Lion City' (1984) there are the lyrics:

> hawker centres on every floor.
> Singapore, Singapore . . .
> Ang Mo Kio – HDB
> Shenton Way – productivity
> People's Park – Keep the city clean . . .
> Singapore, Singapore
> Full of tourists and department stores . . .
> Everything is tall and new and clean . . .
>   (Lee, quoted in Kong 1996: 279)

The fusion of 'traditional market', the drive for economic attainment, the consumerism and control of the society are all apparent. But the whole is set to a Western pop sound and the refrain 'Singapore, Singapore' clearly alludes to 'New York, New York'. Such forms a much more complex relationship of 'the modern', the global and transnational than simple debates over place or placelessness allow (see Chapter 7). Appadurai (1990) suggests that rather than just a singular cultural landscape, we need to see a range of cultural forms and processes coming together in particular local combinations. He suggests studying the conjunctures and disjunctures of ethno-scapes (the cultural map of ethnic identity), media-scapes (the representations of society in various media, see Chapters 5 and 6), ideo-scapes (the range of ideas people have to make sense of the world), techno-scapes (the impact that technologies have in changing relationships over time and space) and finance-scapes (the flows of money and capital at global and local levels). The variable geometry of these different cultural maps produces a fascinating series of intersections – where the differing processes outlined in this book come together to form unique cultural terrains.

ımary

This chapter points to the need to think of cultures and spaces in ways other than as bounded containers. Current studies foreground the more complex cartographies and connections of cultures so that '[t]he very concepts of homogenous national cultures, the consensual or contiguous transmission of historical traditions, or "organic" ethnic communities – *as the grounds of cultural comparativism* – are in a profound state of redefinition' (Bhabha 1994: 5). It may be that cultures are not holistic 'ways of life' – but instead are composed by people assembling and reassembling fragments from around them – the different 'scapes' suggested by Appadurai. Bhabha (1994: 9) suggests that the result of modern history is vast numbers of people 'between' cultures, a *third-space* where extra-territorial and cross-cultural connections create 'unhomely' lives, not rooted in one culture. It is in the juxtaposition, mutations and connections of different cultural spaces, in the overlaying of contradictory cultural landscapes over each other that creativity and vitality may emerge. Such a thirdspace challenges 'our sense of the historical identity of culture as a homogenizing, unifying force, authenticated by an originary Past, kept alive in the national tradition of the People' (Bhabha 1994: 37).

It seems urgent to develop a sense of place that can cope with the globalised world at the moment, where bounded cultures are generally implausible – and the efforts to maintain them taken to dangerous extremes in 'ethnic cleansing'. Removing the sense of bounded culture may be one step to disabling some of the prejudices and dangers that too often form a cornerstone of ethnic nationalism. However, this does not demand an acceptance of a de-territorialised world, or an homogenised type of culture. On the contrary, new cultural forms – neo-tribes – develop codes of belonging, that take up 'traditions' in new and often parodic fashions. So the New Age movement deploys Arthurian fantasy, rediscovered 'tribal wisdom' and the science of chaos and fractal geometry in new and startling combinations. Rave cultures in Britain create spaces that are temporarily outside normal society – spaces where participants can feel emotional belonging, celebrate a culture of the body, dance and freedom in the interstices of Britain's normal cultural landscape. In earlier sections we charted the translation of music over space, but we must also recognise the way it can create spaces of dance, of joy and of transgression (see Chapter 6). We might then look at the *affective spaces* created as moments of emotional change and release.

This is not to support a 'pick'n'mix' version of identity where 'individuals [are] able to select from a plurality of suitably packaged bodies of knowledge in the supermarket of lifestyles' (Featherstone 1991: 112; see Chapter 8). Indeed it is a matter of heated debate whether the commodification of cultures represents a hidden order behind the often chaotic assemblages of modern cultures – an economical imperative of which these forms are only so much cultural clothing. More vexed still is the problem resulting from this. If everyone is immersed in continually shifting and changing cultural landscapes, with different understandings coloured by different situations, then attempting to come up with a single explanation for the 'totality' of events is likely to involve privileging one vantage point – and thus one cultural group – over others. The last chapter of this book thus turns to considering what these understandings of culture mean for how we see academic knowledge.

Further reading

Anderson, B. (1983) *Imagined Communities: Reflections on the Origin and Spread of Nationalism*. Verso, London.

Appadurai, A. (1990) 'Disjuncture and Difference in the Global Cultural Economy', *Theory, Culture & Society* 7: 295–310.

Bhabha, H. (1994) *Nation and Narration*. Routledge, London.

Eade, J. (ed.) (1997) *Living the Global City*. Routledge, London.

Gilroy, P. (1987) *'There Ain't No Black in the Union Jack': The Cultural Politics of Race and Nation*. University of Chicago Press, Chicago.

Gilroy, P. (1993) *The Black Atlantic: Modernity and Double Consciousness*. Harvard University Press, Cambridge, MA.

King, A. (ed.) (1991). *Culture, Globalization and the World System: Contemporary Conditions for the Representation of Identity*. Macmillan, Basingstoke.

Hobsbawm, E. (1990) *Nations and Nationalism since 1780*. Cambridge University Press, Cambridge.

Hosbawm, E. and Ranger, T. (eds) (1983) *The Invention of Tradition*. Cambridge University Press, Cambridge.

Smith, W. (1992) 'Complications of the Common Place: Tea, Sugar and Imperialism', *Jnl of Interdisciplinary History* 13(2): 259–78.

Western, J. (1993) *A Passage to England: Barbadian Londoners Speak of Home*. University of Minnesota Press, Minneapolis.

◗ Cultures of science

translation and knowledge

- ● **The culture of the scientific community**
- ● **Relationships of objective and subjective knowledge**
- ● **Relativism, universal and situated knowledge**

By way of a conclusion, I want to ask *how* we can claim to know things about cultures. This may sound odd after a whole book suggesting different ways of interpreting different forms and practices. However, we have not asked how we might assess whether they are truthful accounts of the world – what is called their *epistemology*. In cultural geography this often raises the ideas of relativism, reflexivity and self-reflexivity. On the first count, *relativism* is often part of the background of cultural study – though not always, and rarely without reservations. Many would regard it as unethical, and often counter-productive, to study a different culture with a view to saying how it is worse than our own or to take our own culture as normal. This does not mean we can never criticise but that we need to be careful that it is not just our prejudices shaping such a criticism. For instance, peoples who live by hunting and gathering may have developed very elaborate cultures – with as many rules and quirks as our own; they may have very sophisticated local knowledges, though they may not have as much technological knowledge. Why call these cultures primitive? A trite example from the developed world would be to attempt to evaluate the culture of a jazz fan with a blues fan – a careful comparison might reveal interesting differences, but saying which is better is likely to prove impossible. This is not to say cultural geographers can never judge. It might be better to say they should be careful never to prejudge.

Equally, geographers do not speak in a silent world, they are one voice among many. Cultural geographers may interpret the transnational links of music say (see Chapters 6 and 10), but this phenomenon has already been interpreted by media (specialist, print, TV), by artists, by listeners, by DJs and by the music industry. There are already multiple interpretations attached to this cultural form before geographers add theirs – people are *reflexive* agents. That is, they already learn from and interpret the world about them as part of normal life. We need some care, then, not only over how we judge different cultures but also about whether we believe our forms of understanding are better than other people's. In short we must be careful not only how we judge between different cultures but also different accounts of the same culture. There are no easy answers, and total relativism would suggest we had nothing worth saying, nothing to contribute – which is about as extreme as saying we always know best. Instead this chapter will briefly look at different approaches to how we assess versions of the world – and suggest how different standards rest upon different assumptions and foundations. The first section will briefly outline some 'traditional' scientific beliefs about what is true. Leading out of this is a critique of the idea that being 'outside' provides better knowledge. I shall suggest that being outside of culture is impossible and what is usually meant is being inside 'scientific culture'. I shall then suggest that the way most cultural geography deals with these problems is to see all knowledge as both partial and situated. This leads to the third idea mentioned above – *self-reflexivity*. This idea is very simple but has profound consequences. At its most basic it suggests that if we cannot get outside culture, if we are always embedded in various value systems just as much as those we study, then we should be scrupulous in examining our presuppositions. Our accounts should acknowledge where we are speaking from as affecting what we say.

Objectivity and knowledge

In Western societies, knowledge has often been structured around binary oppositions – rational against emotional, cultural against natural. A consequence is seeing objective against subjective where the former is privileged. Thus 'objectivity' tends to be valued in knowledge and we find 'that "symbolic" opposes to "real" as fanciful to sober, figurative to literal, obscure to plain, aesthetic to practical, mystical to mundane, and decorative to substantial' (Geertz, quoted in Baker 1993). The question this then raises is how there can possibly be a 'neutral' or objective knowledge about cultures where cultural differences tend to deny an impartial viewing position. The topic of cultural geography is often so 'subjective', about feelings emotions and meanings, that objectivity seems problematic. Some strategies for dealing with this have been covered through the course of this book. For instance, a focus on the material culture of the landscape works to look at how beliefs or meanings are embedded in and expressed through material artefacts (Chapter 2). This is also reflected in the approaches that look to read the landscape in their varied ways – seeing how, for instance, paintings or gardens reflect the cultural beliefs, assumptions of 'lenses' through which the world is viewed (Chapters 3 and 4). Such a concentration on cultural forms thus responds to arguments about 'intangibility' and thus an assumed 'unknowability' of culture.

Other approaches instead embrace and celebrate the idea that human culture is indeed subjective and mysterious in some respects. For instance we have seen the fear that total planning, removing all human foibles from the built environment, is a potentially alienating policy (Chapter 7). Such an approach might also be used to suggest the importance of going beyond 'cultural' objects and seeing how they are inserted in society and daily life. Thus the German Marxist critic Theodor Adorno was working on the consumption of radio programming shortly after the Second World War. The director of the project told him that in order to be scientific, he had to produce some way of measuring, of quantifying, changes in programming and audience reception. Adorno was horrified:

> When I was confronted by the demand to 'measure culture', I reflected that culture might be precisely that condition that excludes a mentality capable of measuring it.
>
> (Cited in Porter 1995: 43)

Adorno (1993) famously criticised the politics of this sort of research. For Adorno, and other thinkers such as Max Horkheimer and Herbert Marcuse of the Frankfurt School, this represented a creeping danger of reducing all life to numbers, that might then form the basis of 'objective' judgements and management. The trend they saw was the development of ever more refined ways of calculating and 'objectively' knowing society that allowed ever more systems of management – both private and public bureaucracies – to dominate people's lives. The result seemed to be that people became objects of knowledge rather than subjects – the power of rationalisations returned to haunt and dominate those in whose service they were meant to be used.

Scientific knowledge in this view is not 'found', truth is not revealed; on the contrary it is constructed. Science, the arts, local belief systems all work to create different knowledges about the world. Saying which one is valid is thus a political issue – it is about empowering the group who sees the world in that way and disabling the arguments of other groups. Differences between different world views may thus be inevitable, as indeed may be the risks of silencing or marginalising groups, but the criteria over which these choices are made are by no means foreordained. The 'objectivity' of science is partial – giving one account of the world – and works to exclude or marginalise other accounts; it is not in that sense neutral. It does not reveal a natural order. If it did there would hardly need to be so many rules governing the conduct of science, nor would studies show that in operation these are generally compromised and often contradictory. An emphasis on the practice of science is important to remind us that we should see people engaged in creating knowledge in different ways. Adorno (1993) highlights that the supposed unification of science through method 'has more to do with administering the world than understanding it. But the bureaucratic imposition of uniform standards and measures has been indispensable for the metamorphosis of local skills into generally valid scientific knowledge' (Porter 1995: 21).

Objective knowledge thus created claims to be universal and untainted by local influences. This claim is extremely important in politically marginalising other forms of knowledge. It can also be enormously effective in regulating social life – something that is not entirely bad in a highly interdependent complex world. This is not an attempt at Luddism, rejecting all scientific knowledge for claiming a spurious objectivity. This kind of knowledge is often invaluable, but a cultural geographer needs to be aware of what such claims to objectivity may imply. Indeed arguing science constructs truths, makes knowledge and is a creative process is not a criticism. It does not make that knowledge wrong or valueless – it is not an attack on science. Instead it is trying to say that we need to rethink how we categorise knowledge, why it is that some knowledge comes to be called universal and others are confined to local belief systems.

Outside cultures: claims to universal truth

One of the tendencies in studying cultures is to equate objectivity with detachment – a separation between observer and observed (see Chapter

7). This is problematic, especially if cultural geography is very often trying to see how people make sense of the world in their own terms by understanding the insider's viewpoint. In one sense we could say that the approaches to iconography (Chapter 3) are caught in precisely this problem – in that they tend to suggest the academic is outside, looking in and saying what is going on from afar, rather than necessarily relating to the actual experiences of people involved in the culture. In part this is inevitable with historical material – even firsthand accounts allow only a vicarious access to the community; all we have are artefacts of varying sorts. However, sometimes this distance means the observer views a culture as a unified whole – 'they', the people studied, are suggested to be similar to each other in contrast to the difference from the researcher. In part this is often true, but there is a balance to be kept in mind. Tuan (1992: 33) argues that people in general try to deal with the messiness of cultures, with the complex patterns of individual characters and beliefs, either by submerging themselves into the group or by attending mostly to commonalities and aspects that suggest order rather than chaos. He concludes that '[a]cademics, who tend to be individualists, favour the second approach: more than other people they seek to escape the world's messiness by withdrawing into a crystalline realm of ideas'. We should be careful, then, of whether 'cultural areas', or indeed holistic, local cultures are not more in the mind of the observer than those of the people they study. Indeed Chapters 5 and 10 suggest we need to be careful indeed about the implications of holistic areas – as to whom they cut out and the implications of boundaries.

The idea of detachment poses some practical problems. A study of subcultures, of (say) gangs or football hooligans, will often be interesting only if it looks at how the people themselves understand what they are doing – how it makes sense to them. But uncritical 'cheerleading' is also unlikely help. The cultural geographer is often in the perilous position of standing between – or, better, shuttling between – different worlds or world-views. The cultural geographer can never stand outside culture. Standing outside the beliefs of those being studied does not mean having no beliefs – instead it means being in your own culture. Many interesting studies have relied on precisely how looking at different cultures reveals the taken-for-granted assumptions of the researcher's own. But neither culture is neutral or objective. There is no Archimedean vantage point from which the cultural geographer can see it as it 'really is'. Street graffiti expressing how gangs see a geography of the city, or academic books saying how geographers see the gangs' territorial patterns are both cultural forms. Neither is neutral or outside. One may circulate more widely around the globe than the other, one may have hugely different effects – no one would deny that, and indeed why one account is adopted or spread instead of another can make the basis of important studies – but both are 'fabrications', ways in which humans have given meaning to their world.

Philosophers, such as Jean-François Lyotard and Ludwig Wittgenstein, have argued that we should thus see the world comprised of various *language games* – that is, ways of describing things and accounting for events that are structured to be internally coherent, and which are acceptable in terms of specific communities. However, these interpretations may well be incommensurable between communities – that is, they may be unintelligible to a different audience, who use different assumptions and rules to decide what is a valid account. These arguments have opened up a huge debate in the social sciences: often described as *postmodern*, they are (to follow Lyotard's definition) hostile to *meta-narratives*. That is, these lines of thought cast doubt on overarching explanations that claim to speak for all people; that claim to be universal and not 'particular'; that claim not to be bounded in a language game unlike the cultures they comment upon. One way in which this debate has developed is to see science not as a finder of universal laws but as a culture in itself.

Cultures of 'outsideness': science and the academy

Much of the emphasis in geography on detachment, or the separation of the observer from the observed as a prerequisite of 'objectivity', can be traced back to the history of the discipline. The idea of a privileged observer as producing truthful knowledge bears the marks of the model of geographical exploration. The relative position of a traveller, moving through territory underpins ideas about an 'outside' view and provides a historical precedent of how knowledge is created in geography. Such travel has the tendency to reduce people encountered to a series of objects   people met in the context of exploration, that is, in relationship to an explorer's journey, not to the context of the rest of their lives or their own ideas of their identity or geography. This then forms a way of looking at the world that turns people into 'objects'. It may well be that this legacy has fed into the way geographers have looked at outside and inside knowledge about cultures.

Cultural geographers have examined the culture of travel, as a way in which geographical knowledge is produced (see Chapter 5). Some researchers have looked at popular travel but, more interestingly for ideas about science, the links of academic knowledge and travel have begun to be explored. Looking at the practices of travellers, researchers have found not so much a detachment but an active repression of evidence of contact. Thus writing in the passive voice to describe people and landscapes (e.g. 'the river was crossed') denies any sense of agency – and, given that many explorers had large numbers of local people as porters, represses the copresence of explorer and people. Likewise talking of peoples in a generic term serves to make them silent objects of study not groups of people with whom the researcher interacted (famously the anthropologist Evans Pritchard referred to 'the Nuer', never individuals whom he met). Amid all this the ethnic divisions and relations of power and wealth that enabled the explorer to travel tend to be downplayed. Exploration was rarely disinterested; newspapers funded missions to generate sensational stories; colonial powers sought new markets or resources; and military planners promoted the idea of geographical training as useful to empire. In addition the embodied nature of the explorer, their gendered position tends to be omitted. It is clear from 'adventure stories' that the heroic explorer on the frontier of civilisation and knowledge was set up as a romantic ideal for boys to aspire to – travel 'alone' in foreign lands was set up in the image of a particular sort of masculine identity. In short, it has been argued that the impression of a detached, objective knowledge is more a textual, rhetorical one – an appearance created through conventions of how accounts were written up – rather than an accurate model of how knowledge was created.

So much for the history of exploration, but does this impact on other scientific modes of looking at the world? One lesson is to study the practices of 'knowing' rather than just the accounts of what is known. We cannot say knowledge is ever fully detached from the locales where it is created. It circulates through academic institutions and learned societies designed to allow the exchange of knowledge – it does not float free but relies on these knowledge-producing networks. Transmission does not mean researchers are detached, rather that others are learning the tacit knowledge and assumptions necessary to understand the new research. Even in the most rigorous science, based in laboratories working on DNA or physics, we can suggest that the idea of mechanistic objectivity, where knowledge is based completely on explicit rules, is never fully attainable. Even in the physical sciences the importance of tacit knowledge is now widely recognised. Thus let us think of something as 'objective' as sheer strain in soil. Currently one of my colleagues has one of only seven machines in the country capable of performing a particular test on soil samples. Obviously then transmitting and developing the ideas derived from experiments will mean transmitting practical skills to others about how the machinery works:

> Experimental success is reflected in the instruments and methods as well as factual assumptions of other laboratories. Day-to-day science is as much about the transmission of skills and practices as about the establishment of theoretical doctrines.
>
> (Porter 1995: 12)

If that applies to experimental science it applies as much to assessing knowledge about cultures. We may well have then to look at the moral economy of science as a culture – a culture that rewards and values diligence and that relies on trust in respecting others' ideas. It is in that sense a culture about knowledge, where the value of ideas tends to be defined by other researchers. That is, ideas are not held up in isolation but are assessed by a community of fellow experts. The ideas, be they about soil or culture, are then judged according to the norms of that community – using tacit knowledge, practical experience and so forth to assess the worth of any contribution.

Situated knowledge

Cultural geographies are embedded in a series of relationships. There is, first, the relationship with the people studied but, second, there is the position within the academy. So many would argue that there is no kernel of absolute truth for all time – we cannot bracket out 'impurities'. It is not the case that social factors (our backgrounds, the context of research) can either be factored out, or that they devalue our knowledge; instead such tacit or practical factors are vital in creating knowledge. They cannot then be simply removed as though they contaminate or corrupt the work. Scientific knowledge should not be seen as being contaminated or 'biased' by social factors; rather science should be seen as a social process.

The logic of this is to say that cultural geography should not be in the business of creating absolute truths – as though they were true for all people – because there is no position where such detached, disembedded and asocial knowledge could be created or circulated. Knowledge, academic or popular, is about cultural systems of belief and validation – and cultural geography does not escape that. So how can we see ways forward? One way is to say this highlights a need to think about whom and how we are studying. Anthropologists have recently remarked that in the study of the cultures of the globe, modern Western society is often absent, as though it did not have a culture(s). We need then to think that cultural geography is not just a matter of studying exotic other peoples, but thinking about how we define them as 'exotic', what is thus going on in our own taken-for-granted worlds.

The situatedness of knowledge in and between the cultures of researcher and researched highlights the importance of thinking about why we carry certain assumptions and connecting our biographies to what we study. This is generally said to be about *self-reflexivity* and is marked in the very least by using the first person – saying what you and others did rather than hiding it in a passive voice. Going further it is generally marked by an attention to tacit assumptions made by the researcher (often requiring a fairly hard, long self-analysis) or made about the researcher. It is thus very much attuned to the social process of creating knowledge. Of course there is a problem in all this that such careful self-reflection, such thinking about the research process, may downplay the original aim of the research. These works are also generally marked by an attention to writing. That is, they do not see writing as passively transmitting information but as playing an active part in constructing an idea of the world for the readers. The idea of texts as transparently reflecting reality is as much a rhetorical strategy as any other style of writing. The common academic voice of the impersonal narrator distances us from what is recounted making it appear self-evident, repressing the activity that went into producing the account, whom it allows to speak and whom it may silence.

Current work suggests we need to examine this process, to look at the way writing creates particular effects. This concern for the process of shaping and transmitting knowledge suggests that one way of thinking of cultural geography is as 'a translation', a making of connections between different ways of seeing the world. Rather than seeing our position between different interpretive frames – those of our own culture, academia, and the culture we study – as constituting a problem, we can think of it as the most exciting and exhilarating place of all. In a world of increasingly rapid change and flows, such points of contact are becoming all the more common between groups and cultures. Cultural geography may thus be one of the best avenues through which to address these changing definitions of who is an insider and an outsider, who knows what about whom and how we adapt to new ways of being in the world.

Summary

This chapter has not sought to bring the studies in previous chapters to a conclusion. It does not resolve the differences between them or sum them up into a grand pattern. Instead it tries to leave a few questions that lead on to issues of greater depth. It is concerned with the practice and process of academic knowledge. It has tried to point out how we see academic accounts producing truthful knowledge. It has asked how we set up criteria to judge this, and has made an argument that we need to be sensitive to cultural difference in such judgements. This is especially important given both the cultural topic of study and the legacy of colonial geography; claims to objective, absolute knowledge have been closely linked to real exploitation and colonisation (see also Chapter 5). I have thus argued we need to be aware of our own position in producing knowledge, seeing it as a process of actively making academic knowledge rather than finding already-existing truths. The model then for the cultural geographer may be that of a translator or intermediary rather than an arbiter of what is right and wrong. It will be clear that the different approaches in different chapters react to these issues in different ways, answering the challenges or disputing them according to their own lights. It is my hope that as you approach these different topics in more detail in later years these questions of how we can claim to know things, and the implications of this will continue to develop.

Further reading

Barnes, T. (1996) *Logics of Dislocation: Models, Metaphors and Meanings of Economic Space* (esp. chs 4 and 5). Guilford Press, New York.

Bryant, R. (1996) 'Romancing Colonial Forestry: The Discourse of Forestry as Progress in British Burma', *The Geographical Journal* 162(2):169–78.

Clifford, J. and Marcus, G. (eds) (1986) *Writing Culture: The Poetics and Politics of Ethnography*. University of California Press, Berkeley.

Porter, T. (1995) *Trust in Numbers: The Pursuit of Objectivity in Public Life*. Princeton University Press, Princeton, NJ.

Bondi, L. and Domosh, M. (1992) 'Other Figures in Other Places: On Feminism, Postmodernism and Geography', *Society and Space* 10: 199–213.

Duncan, J. and Ley, D. (eds) (1992) *Place/Culture/Representation*. Routledge, London.

Riffenburgh, B. (1993) *The Myth of the Explorer*. Oxford University Press, Oxford.

 **Glossary**

The following can only be the barest of outlines, or signposts, if you will. The items below are not really definitions so much as comments aimed at the way these terms have been used in cultural geography. I hope they are helpful in clarifying some of the terms used in this book and in the further reading suggested. I also hope that, in due course, the inadequacies of these summaries will become apparent and prompt further thought about what terms may really mean in specific contexts. The terms here are ones which my own tutees (mostly) have asked for help with on varying occasions, otherwise the list could of course be endless. One of the reasons for all the jargon is that in order to think differently, to open up new possibilities, it is often necessary to think in new categories or languages. Sometimes they can also be a form of shorthand used to encapsulate long arguments in short phrases – rather like using a label. Many more detailed and erudite descriptions, though not always easy ones, can be found in the *Dictionary of Human Geography*. As a reading tactic, given the pressures on your time, I suggest that if a word baffles in a book, you 'beep' over it, read the whole paragraph and see if its sense does not become clearer – it often does. If it really is a sticking point then go and look it up.

Constructivist An approach that suggests we cannot directly access the world, and the categories and things we see are shaped by our interpretations rather more than the world outside. Indeed strong forms of this theory argue we can never know 'the outside', as the American philosopher Rorty put it, it is interpretation all the way down.

Contingent A term in logic that suggests the connections between things or events are not NECESSARY but are the outcome of historical and geographical context (they are thus also often called 'contextual'). Thus a pattern of events or agents may cause something to happen though none of them is necessary for it to do so. This means that the same causes are not always associated with the same event, and that the same sort of events can be brought about by different causes.

Deconstruction   An approach borrowed from literary theory and identified with the work of Jacques Derrida. It developed from ideas about SIGNIFICATION to argue that meanings in texts, be they written or visual, are not fixed but open to play. Thus any story will have exclusions, things it cuts out, which form constitutive absences. These absences allow it to make sense. However, they form a sort of ghostly mark and, like the ghost in Hamlet, they drive the plot. Deconstruction aims to bring these sorts of exclusions to light while also opening up connotations and implications in texts. It thus suggests that representation can never be simple or transparent. There is no neutral language to describe the world. This has, to say the least, been an extremely influential and controversial argument.

Determinism   Approaches which seek to find a single cause (or more rarely a small number) for events are called determinist. These approaches tend thus to simplify and reduce (and are often called 'reductionist') events to key causal factors. Given that most events have a surfeit of agents and forces shaping them this normally means prioritising one or suggesting that it lies behind other events. See also NECESSARY and CONTINGENT.

Dialectic   This term refers to the relationships between ideas or things. In its strictest sense there are three parts to a dialectic: a thesis, that is, the first idea or trend; its anti-thesis, which is an opposite or antagonistic reaction caused by the thesis; the interaction of these two leads to a synthesis that resolves the contradictions between them. Often the last term is not used, and dialectical comes to mean one process causing two outcomes that are paradoxical or antagonistic to each other. Note how this means the conflict drives ideas and things forwards towards a resolution or crisis – dialectics imply things are developing not standing still.

Diaspora   Originally derived from accounts of the Jewish people outside Palestine. This term has been used to try to evoke a modern condition where belonging is not fixed according to territorial possession, where different peoples mingle one among the other. It also, though, evokes a sense of exile and homelessness that echoes experience in the more mobile modern societies of the present.

Discourse   Many of you will be familiar with arguments about how people's geographical perceptions influence their actions. Normally this is described in terms of deviation from some real geographical world. The term discourse suggests that representations and images are constitutive of the world, and what is 'out there' is actively shaped by ideas and concepts. Discourse is a word applied first to the way we talk about the world, the figures of speech we use. At this level it implies that the way we discuss issues may also impact on how we act and how those issues are re-produced. Second, discourse can apply to non-verbal processes, such as the pictures and images used to portray events and places. It is difficult to think how we would recognise a place or an event without some combination of the above two processes – even scientists never see reality, they can only ever see it through their theories.

Dualism A form of argument where things encountered are divided into two categories and have to fall into one or the other. This means that the two categories are not seen as mixed or combined. A much discussed example is the 'mind–body' dualism where rational thought is opposed to irrational bodily desires. This plays a crucial role in Chapter 5. A similar process underpins the ideas of 'dichotomy'.

Enlightenment Project A philosophical and intellectual project generally dated to start in the seventeenth and eighteenth centuries. It implies a change in outlook on the world where science was seen as the key to demystifying the world, and the world was believed to be explicable by rational thought. The processes of the world were now seen as coherent and knowable rather than religious and mysterious. This form of thinking has been criticised for emphasising the rational mind as though it were free from social and material entanglements. So although Enlightenment projects spoke of universal truths they were often based on the experience of white, male European scientists.

Empiricism A form of study that concentrates its attention on the visible or observable. Such studies have little concern with theory generation but seek to look at facts. They are then often criticised for assuming that facts occur independently of how they are observed and are 'theory–neutral'.

Essentialism This is applied to interpretations that seek to ascribe a fundamental and unchanging core value to something. These interpretations look for things that of themselves define a group or type of event. These core elements are not then seen as negotiable or contestable.

Gender Most cultural geographers do not simply equate sex, linked to a person's anatomy, to gender. In Simone de Beauvoir's phrase they suggest gender is learnt – it is seen as sets of expectations and appropriate behaviour which vary over space and time. Not only that but they may comprise different ways of using space and conceptions of space. Thus studies will look at femininity or masculinity as cultural identities, that are place-specific and entail different spatial behaviours.

Hegemony A term derived from the Italian Marxist Antonio Gramsci. Broadly put it involves the idea that the ruling groups in society have to 'manufacture consent' to their dominance. At this point many people disagree over how and the extent to which this is done. The theme I wish to draw out is that Gramsci first suggested that in order for the dominant class to maintain economic dominance it had to establish legitimating or supporting structures in the social and cultural reproduction of society. He thus suggested the importance of contesting and resisting dominant ideas and concepts (although he never used the term 'discourses') as well as simply economic struggle.

Hermeneutics This is a term for a process of textual interpretation. Broadly it suggests that the scientific method is miscast and that, rather than seeking a neutral and authoritative knowledge, the normal and real process is one of conversation. This is where two systems of knowledge, say, you as researcher and the people you research, come into contact. Both can say the same words and mean different things, both have different understandings of the world. In a process of dialogue they generate some form of shared understanding.

Iconography Closely related to SEMIOTICS but tending to look at signs embedded in the landscape or material artefacts. Thus it might attempt to read the symbolism of the urban fabric, or what the landscape can say about the values and society of the people who shaped it. In geography a great deal of attention has focused on how the landscape represents and embodies patterns of power, dominance and HEGEMONY and how this is in turn represented in art, pictures and descriptions. It has been argued that we may treat landscape as a text that thus comes to embody a range of discourses that affect how people act.

Idealist Forms of explanation that tend to locate the most important elements in mental or symbolic categories. In cultural geography this tends to mean looking at meanings in culture rather than artefacts, material conditions and so on. Often opposed to MATERIALISM and often used as a criticism of CONSTRUCTIVIST approaches.

Idiographic A form of study that emphasises the individual and particular. In terms of geography it has been associated with studies that emphasise places as unique assemblages where diverse processes and factors come together. In such cases it is often argued that geography should be the study and detailed accounting of these places – literally a writing of the particular local circumstances. Associated with a background in 'regional geography', the idiographic tradition tended to study areal differentiation (in Hartshorne's phrase). It is often counterposed to the NOMOTHETIC tradition.

Materialist Forms of explanation that tend to locate the most important elements in material factors rather ideas and symbols. Thus Marx criticised the German philosopher Hegel for being IDEALIST by saying he mistook the things of logic for the logic of things.

Necessary A term in logic that means something cannot happen or exist without this element being present. It should be distinguished from 'sufficient'; something can be necessary for an event to happen without being sufficient on its own to make it happen. See also CONTINGENT.

Nomothetic A form of study that emphasises spatial regularities and patterns. It has often been associated with attempts to derive laws and predict patterns – thus to find rules that if x and y occur together z will happen. The approach is often associated with the quantitative revolution in geography and the rise of the scientific method. The focus is on spatial process rather than particular circumstances, which are seen as deviations or contextual effects. Often counterposed to IDIOGRAPHIC approaches.

Palimpsest A term deriving from medieval writing. A writing tablet would be used and then reused for a new set of material. However, the tablet could never be entirely rubbed clean. So over time and with successive reuses layers of prior scripts would build up over which the current one was written. The term has thus been taken as a metaphor for the processes of landscape change, where current uses over-write but do not completely erase the marks of prior use.

Polity The term refers to a political entity, that is, a state defined through its mode of government. Thus one has democratic polities, authoritarian polities and so forth. At a more fine-grained level many would say the idiosyncrasies of particular nations inflect their political systems, so within the broad categories you might see 'British polity' implying the particular systems and formations developed in the UK.

Race and ethnicity These are not regarded as biological categories. Chapters 5 and 11 argued that much of the science of classifying people by race was driven by Western (and white) ideological concerns. It remains difficult to address the lived reality that everyone is born with a colour while not reinforcing the idea that our interpretations of these colours is natural.

Reflexivity This term reminds us that people learn, so they do not simply repeat actions but learn from the experience of last time. Society is thus dynamic because people learn the rules it follows, thus changing the outcomes. Take a simple example: if a tutor gave more marks to typed pieces of work soon everyone would type and thus any initial advantage for those who already typed essays would soon disappear. This is often compounded by issues of SELF-REFLEXIVITY, for which it is sometimes used as a shorthand.

Self-reflexivity If we attend seriously to the critiques of neutral knowledge, then we have to ask what are our own motivations, cultural baggage and inclinations and how they affect our studies. So self-reflexivity means being aware that research is not looking at 'them' and 'their' culture, but at our cultural background embodied in us, meeting their cultural background, embodied in them, in the research process.

Semiotics A term meaning the study of signs derived from formal and structuralist linguistics. Semiotics is thus an attempt to read the meanings of the world through studying what languages and symbols occur. Given its roots the way it has traditionally attempted to do this is looking for patterns and regularities in the relationships *between* signs. The signs may be actually in texts – that is, written, spoken, in media or constructed in any form. The study of semiotics emphasises the processes of communication.

Sign, Signifier, Signified The world is replete with signs which produce meanings. At the most basic levels signs stand in for something else: thus a word stands in for a concept, an illustration on a road-sign of someone digging stands in for a picture of the actual work. From the work of the linguist Ferdinand de Saussure, many divide signs into signifiers and signifieds. Signifieds are what are referred to – the objects, as it were. The signifiers are the things that refer to them. However, a great deal has been made of the possible slippages between the two, where there is no necessary linkage of signifier to signified. The example Saussure started from was catching the 8.45 train to Nancy (the sign); in that case very few of us would recognise the train (the signified) unless we read the locomotive numbers or something (though that only postpones the problem). Rather we would know it was the 8.45 to Nancy because it was there at 8.45 not 8.00 (being French railways there is a hope it would be on time) and because it was going to Nancy not Montpellier. In other words we know what it is through a relation between signifiers, through what it is not, instead of through a link to what is signified. The whole process of thus creating meaning is called 'signification'.

Teleology   A directed story where the predetermined end-point guides the interpretation of prior events. Derived originally from theology it applies to theories that posit a generally historical evolution towards an inevitable end-state.

Time–Space, or **Space–Time**   These terms came into geography through work on 'time geography' pioneered by Hagerstrand. At a simple level we might think of them as introducing a third dimension to maps. Thus as well as x and y coordinates there is a time axis. On a daily basis we look at how our use of space is structured by time – looking at how different groups coalesce or disperse at different locations and times. On a deeper level it suggests an attention to the everyday and routine conditions of life can reveal many deep-seated patterns in society. Geographers have thus looked at the space–time envelopes that different sorts of people have to look at inequalities in life.

References

Abbeele, G. Van der (1991) *Travel as Metaphor: From Montaigne to Rousseau.* University of Minnesota Press, Minneapolis.

Adorno, T. and Horkheimer, M. (1947) *The Dialectic of the Enlightenment.* Verso, London.

—— (1991) *The Culture Industry: Selected Essays.* Routledge, London.

—— (1993) 'Messages in a Bottle', *New Left Review* 200: 5–14.

Alvarez, A. (1995) *Night: NightLife, Night Language, Sleep and Dreams.* Norton, New York.

Anderson, B. (1983) *Imagined Communities: Reflections on the Origin and Spread of Nationalism.* Verso, London.

Anderson, P. (1990) 'A Culture in Contraflow', *New Left Review* 180: 41–80.

Appadurai, A. (1990) 'Disjuncture and Difference in the Global Cultural Economy', *Theory, Culture & Society* 7: 295–310.

Augé, M. (1995) *Non-Places: Introduction to an Anthropology of Supermodernity.* Verso, London.

Baker, S. (1993) *Picturing the Beast: Animals, Identity and Representation.* Manchester University Press, Manchester.

Baudrillard, J. (1989) *America.* Verso, London.

Bauman, Z. (1992) 'Soil, Blood and Identity', *Sociological Review*: 675–701.

Benjamin, W. (1973) *Illuminations*, trans. H. Zohn. Fontana, London.

—— (1974) *Charles Baudelaire: Lyric Poet in the Era of High Capitalism.* New Left Books, London.

Barnett, A. (1990) 'Cambodia Will Never Disappear', *New Left Review* 180: 101–26.

Bennett, T. (1988) 'The Exhibitionary Complex', *New Formations* 4: 73–102.

Berman, M. (1983) *All That Is Solid Melts into Air.* Verso, London.

Beynon, H. (1973) *Working for Ford.* Allen Lane, London.

Bhabha, H. (1994) *Nation and Narration.* Routledge, London.

Bianchini, F. (1993) 'Culture and the Remaking of European Cities', pp. 1–20 in Bianchini, F. and Parkinson, M. (eds), *Cultural Policy and Urban Regeneration.* Manchester University Press, Manchester.

Blunt, A (1994) *Travel, Gender and Imperialism: Mary Kingsley in West Africa.* Westview Press, Boulder.

Blunt, A. and Rose, G. (eds) (1995) *Writing Women and Space: Colonial and Postcolonial Geographies*. Guilford Press, New York.

Bondi, L and Domosh, M. (1992) 'Other Figures in Other places: On Feminism, Postmodernism and Geography', *Society and Space* 10: 199–213.

Bourdieu, P. (1984) *Distinction : A Social Critique of the Judgement of Taste*, trans. R. Nice. Routledge, London.

—— (1990) *Logic of Practice*. Polity Press, Cambridge.

—— (1991) *The Political Ontology of Martin Heidegger*. Polity Press, Cambridge.

—— (1995) *The Field of Cultural Production*. Polity Press, Cambridge.

Bowlby, R. (1985) *Just Looking: Consumer Culture in Dreiser, Gissing and Zola*. Methuen, London.

Boyes, G. (1995) *The Imagined Village: Culture, Ideology and the English Folk Revival*. Manchester University Press, Manchester.

Brantlinger, P. (1985) 'Victorians and Africans: The Genealogy of the Myth of the Dark Continent. *Critical Inquiry* 12: 166–203.

—— (1993) *Rule of Darkness: British Literature and Imperialism, 1830–1914*. Cornell University Press, Ithaca.

Brantlinger, P. and Naremore, J. (eds) (1991) *Modernity and Mass Culture*. Indiana University Press, Bloomington.

Brosseau, M. (1995) 'The City in Textual Form: *Manhattan Transfer's* New York', *Ecumene* 2(1): 89–114.

Buck-Morss, S. (1989) *The Dialectics of Seeing: Walter Benjamin and the Arcades Project*. MIT Press, Cambridge, MA.

—— (1986) 'The Flâneur, the Sandwichman and the Whore: The Politics of Loitering', *New German Critique* 39: 99–139.

Campbell, B. (1992) *Goliath: Britain's Dangerous Places*. Methuen, London.

Castells, M. (1989) *The Informational City*. Blackwell, Oxford.

Certeau, M. de (1984) *The Practice of Everyday Life*. California University Press, Berkeley, CA.

—— (1988) *The Writing of History*, trans. T. Conley. Columbia University Press, New York.

Cockburn, C. (1985) *Brothers: Male Dominance and Technological Change*. Pluto, London.

Coleman, A. (1985) *Utopia on Trial: Vision and Reality in Planned Housing*. Shipman, London.

Collins, J. (1996) *Architectures of Excess*. Verso, London.

Collier, P. (1991) 'The Inorganic Body and the Ambiguity of Freedom', *Radical Philosophy* 57: 3–9.

Cook, I. (1995) 'A Grumpy Thesis', PhD thesis submitted to the University of Bristol.

—— (1996) 'Tropics of Consumption: Representing Exotic Fruits in British Culinary Culture'. Mimeo.

Cook, I. and Crang, P. (1996) 'The world on a Plate – Culinary Culture, Displacement and Geographical Knowledges', *Journal of Material Culture* 1(2): 131–53.

Corbin, D. (1981) *Life, Work and Rebellion in the Coal Fields: The Southern West Virginia Miners 1880–1922*. University of Illinois Press, Urbana.

Coupland, D. (1994) *Microserfs*. Fontana, London.

Crang, M. (1996) 'Envisioning Urban Histories: Bristol as Palimpsest, Postcards, and Snapshots', *Environment and Planning* A 28 3: 429–52.

Crang, P. (1995) 'It's Showtime': on the Workplace Geographies of Display in a Restaurant in Southeast England', *Society and Space* 12(6): 675–704.

—— (1996) 'Displacements: Geographies of Consumption. *Environment and Planning A* 28(1): 47–68.

Cresswell, T. (1993) Mobility as Resistance: A Geographical Reading of Kerouac's *On the Road, Trans. Inst. Br. Geogr.* (NS) 18: 249–62.

Daniels, S. (1993) *Fields of Vision: Landscape Imagery and National Identity in England and the US*. Polity Press, Cambridge.

Daniels, S. and Rycroft S. (1993) 'Mapping the Modern City: Alan Sillitoe's Nottingham Novels', *Transactions of the Institute of British Geographers* 18(4): 460–80,

Darby, H. C. (1948) 'The Regional Geography of Thomas Hardy's Wessex', *Geographical Review* 38: 426–43.

Davis, M. (1990) *City of Quarz: Excavating the Future in Los Angeles*. Verso, London.

Dayan, D. and Katz, E. (1985) 'Electronic Ceremonies: Television Performs a Royal Wedding', pp. 16–32 in Blonsky, M. (ed.), *On Signs*. Blackwell, Oxford.

Deleuze, G. and Guattari, F. (1987) *A Thousand Plateaux*. University of Minnesota Press, Minneapolis.

Deoliver, M. (1996) 'Historical Preservation and Identity – The Alamo and the Production of a Consumer Landscape', *Antipode* 28(1): 1–20.

Didion, J. (1979) *Run River, Run*. Harmondsworth, Penguin.

Dohse, K. Jurgens, U. and Malsch, T. (1985) From "Fordism" to "Toyotism"? The Social Organisation of the Labour Process in the Japanese Automobile Industry', *Politics and Society* 14(2).

Donald, J. (1993) 'How English Is It? Popular Literature and National Culture', pp. 165–86 in Carter, E., Donald, J. and Squires J *Space and Place: Theories of Identity and Location*. Lawrence & Wishart: London.

Douglas, M. and Isherwood, B (1978) *The World of Goods. Towards an Anthropology of Consumption*. Allen Lane, London.

Duncan, J. (1981) 'The Superorganic in American Cultural Geography', *Annals Assoc. Amer. Geogr.* 70: 181–92.

—— (1990) *The City as Text: The Politics of Landscape Interpretation in the Kandyan Kingdom*. Cambridge University Press, Cambridge.

Duncan, J. and Ley, D. (eds) (1992) *Place/Culture/Representation*. Routledge, London.

Dundes, A. (1985) Nationalistic Inferiority Complexes and the Fabrication of Fakelore: A Reconciliation of *Ossian*, the *Kinder- und Hausmärchen*, the *Kalevala*, and Paul Bunyan', *Journal of Folklore Research* 22(1): 5–18.

Eco, U. (1987) *Travels in Hyper-reality*. Picador, London.

Eisenstein, S. (1943) *The Film Sense*. Faber, London.

Enloe, C. (1989) *Bananas, Beaches and Bases: Making Feminist Sense of International Politics*. University of California Press, Berkeley.

Eyerman, R. and Löfgren, O. (1995) 'Romancing the Road: Road Movies and Images of Mobility', *Theory, Culture & Society* 12: 53–79.

Featherstone, M. (1991) *Consumer Culture and Postmodernism*. Sage, London.

Ferguson, P. (1994) 'The Flâneur on and off the Streets of Paris', in Tester, K. (ed.), *The Flâneur*. Routledge, London.

Ford, L. (1994) 'Sunshine and Shadow: Lighting and Color in the Depiction of Cities in Film, pp. 119–36 in Aitken, S. and Zonn, L. (eds), *Place, Power, Situation and Spectacle: A Geography of Film*. Rowman & Littlefield, Lanham.

Forêt, P. (1995) 'The Manchu Landscape Enterprise: Political, Geomantic and Cosmological Readings of the Gardens of the Bishu Shanzhuang Imperial Residence at Chengde', *Ecumene* 2(3): 325–34.

Frisby, D. (1992) *Fragments of Modernity*. Sage, London.

Frobel, F., Heinrich, J. and Kreye, O. (1980) *The New International Division of Labour*. Cambridge University Press, Cambridge.

Fucini, J. and Fucini, S. (1990) *Working for the Japanese: Inside Mazda's American Auto Plant*. Free Press, Toronto.

Gill, A. (1995) *Ruling Passions: Sex, Race and Empire*. BBC Books, London.

Gilman, S. (1985) 'Black Bodies, White Bodies: Toward an Iconography of Female Sexuality in Late Nineteenth-Century Art, Medicine and Literature', *Critical Inquiry* 12(1): 223–61.

Gilroy, P. (1993) *The Black Atlantic: Modernity and Double Consciousness*. Harvard University Press, Cambridge, MA.

Gold, J. (1985) 'From *Metropolis* to *The City*: Film Visions of the Future City, 1919–39', pp. 123–38 in Burgess, J. and Gold, J. (eds), *Geography, the Media and Popular Culture*. Croom Helm, London.

Gold, J. and Ward, S. (1994) 'We're Going to Do it Right This Time: Cinematic Representations of Urban Planning and the British New Towns, 1939–51', pp. 229–58 in Aitken, S. and Zonn, L. (eds), *Place, Power, Situation and Spectacle: A Geography of Film*. Rowman & Littlefield, Lanham.

Goss, J. (1993) 'The Magic of the Mall', *Annals of the Association of American Geographers* 83: 18–47.

Gould, S. (1994) 'American Polygeny and Craniometry before Darwin: Blacks and Indians as Separate Inferior Species', pp. 84–115 in Harding, S. (ed.), *The 'Racial' Economy of Science*. Indiana University Press, Bloomington.

Grace, E. (1990) *Shortcircuiting Labour: Unionising Electronics workers in Malaysia*. INSAN, Kuala Lumpur.

Greenwood, D. (1977) Culture by the Pound: An Anthropological Perspective on Tourism as Cultural Commoditization', pp. 129–38 in Smith, V. (ed.), *Hosts and Guests: The Anthropology of Tourism*. University of Pennsylvania Press, Philadelphia.

Gregory, D. (1994) *Geographical Imaginations*. Basil Blackwell, Oxford.

Gregory, D. (1991) 'Interventions in the Historical Geography of Modernity: Social Theory, Spatiality and the Politics of Representation', *Geografiska Annaler* 73 (B) 1: 17–44.

Grossman, R. (1979) 'Women's Place in the Integrated Circuit', *Pacific Research/South East Asian Chronicle*, Special issue.

Grunnwald, J. and Flamm, K. (1985) *The Global Factory: Foreign Assembly in International Trade*. Brookings Institution, Washington DC.

Haggard, R. (1885) *King Solomon's Mines*. (Reprinted 1982, Ladybird, London.)

Hall, S. (1991) 'Old and New Identities, Old and New Ethnicities', pp. 41–68 in King, A. (ed.), *Culture, Globalization and the World System*. Macmillan, Basingstoke.

Handler, R. and Linnekin, J. (1984) 'Tradition, Genuine or Spurious', *Journal of American Folklore* 97 (385): 273–290.

Harvey, D. (1985) *Urbanisation of Consciousness*. Blackwell, Oxford.

—— (1989) *The Condition of Postmodernity*. Blackwell, Oxford.

—— (1993) 'From Space to Place to Back Again' in J. Bird *et al.* (eds), *Mapping the Futures*. Routledge, London.

Haug, W. (1987) *Commodity Aesthetics: Ideology and Culture*. International General, New York.

Helms, M. (1989) *Ulysses Sail: Travel, Knowledge and Power*. Princeton University Press, Princeton, NJ.

Hinsley, C. (1991) The World as Market Place: Commodification of the Exotic at the World's Columbian Exposition, Chicago 1893', pp. 344–65 in Karp, I. and Lavine, S. (eds), *Exhibiting Cultures: The Poetics and Politics of Museum Displays*. Smithsonian Press, Washington, DC.

Hobsbawm, E. (1990) *Nations and Nationalism*. Cambridge University Press, Cambridge.

Hobsbawm, E. and Ranger, T. (eds) (1989) *The Invention of Tradition*. Cambridge University Press, Cambridge.

Holquist, M. (1990) *Dialogism: Bakhtin and his World*. Routledge, London

Hoskins, W. (1955) *The Making of the English Landscape*. Penguin, London.

Jackson, P. (1989) *Maps of Meaning*. Routledge, London.

—— (1995) 'Manufacturing Meaning: Culture, Capital and Change', pp. 165–89 in Rogers, A. and Vertovec, S. (eds), *The Urban Context: Ethnicity, Social Networks, and Situational Analysis*. Berg, Oxford.

Johnson, N. (1995) 'Cast in Stone: Monuments, Geography, and Nationalism', *Society and Space* 13: 51–65.

Kaarsholm, P. (1989) 'The Past as Battlefield in Rhodesia and Zimbabwe', *Culture and History* 6: 85–106.

Kern, S. (1983) *The Culture of Time and Space*. Cambridge University Press, Cambridge.

Kong, L. (1996) 'Popular Music in Singapore: Exploring Local Cultures, Global Resources and Regional Identities', *Society and Space* 14: 273–92.

Lacan, J. (1977) *Écrits: A Selection*. Routledge, London.

Lears, T. (1989) *Fables of Abundance*. Basic Books, New York.

Leed, E. (1991) *The Mind of the Traveller: From Gilgamesh to Global Tourism*. Basic Books, New York.

Leidner, R. (1993) *Fast Food and Fast Talk: Service Work and the Routinization of Everyday Life*. University of California Press, Berkeley, CA.

Leppert, R. (1993) *The Sight of Sound: Music, Presentation and the History of the Body*. University of California Press, Berkeley, CA.

Lester, E. (1992) 'Buying the Exotic Other: Reading the 'Banana Republic' Mail Order Catalog', *Jnl of Communication Inquiry* 16(2): 74–85.

Lewis, P. (1987) 'Taking Down the Velvet Rope: Cultural Geography and the Human Landscape', pp. 23–9 in Blatti, J. (ed.), *Past Meets Present: Essays about Historic Interpretation and Public Audiences*. Smithsonian Institute Press, Washington, DC.

Ley, D. and Olds, (1988) 'Landscape as Spectacle: World's Fairs and the Culture of Heroic Consumption', *Society and Space* 6: 191–212.

Ley, D. and Samuels, M. (1979) *Humanistic Geography*. Croom Helm, London.

Leyshon, A., Matless, D. and Revill, G. (1995) 'The Place of Music', *Transactions of the Institute of British Geographers* 20: 423–33.

Linebaugh, P. and Rediker, M. (1990) 'The Many-headed Hydra: Sailors, Slaves and the Atlantic Working Class in the Eighteenth Century', *Jnl of Historical Sociology* 3(3): 225–52.

Lonsdale, J. (1992) 'African Pasts in African Future', *Canadian Journal of African Studies* 23: 126–46.

Low, G. C.-L. (1993) 'His Stories? Narratives and Images of Imperialism', pp. 187–220 in Carter, E., Donald, J. and Squires, J. *Space and Place: Theories of Identity and Location*. Lawrence & Wishart, London.

Lowenthal, D. (1984) *The Past Is a Foreign Country*. Cambridge University Press, Cambridge.

Lyotard, J.F. (1984) *The Postmodern Condition: A Report On Knowledge*. Manchester University Press, Manchester.

MacCannell, D. (1976) *The Tourist: A New Theory of the Leisure Class*. Shocken, New York.

—— (1992) *Empty Meeting Grounds: The Tourist Papers*. Routledge, London.

Macdonald, G. (1995) 'Indonesia Medan-Merdeka – National Identity and the Built Environment', *Antipode* 27(3): 270–93.

Macdonald, R. (1993) *Sons of the Empire: The Frontier and the Boy Scout Movement, 1890–1914*. University of Toronto Press, Toronto.

Maffesoli, M. (1996) *The Time of the Tribes*. Sage, London.

Marcuse, H. (1964) *One-Dimensional Man*. Routledge, London.

Massey, D. (1994) *Space, Place and Gender*. Polity Press, Cambridge.

Massey, D., Quintas, and Wield, (1992) *High-Tech Fantasies*. Routledge, London.

Matless, D. (1993) 'One Man's England: W.G. Hoskins and the English Culture of Landscape', *Rural History* 4(2): 187–207.

McCracken, G. (1990) *Culture and Consumption: New Approaches to the Symbolic Character of Consumer Goods and Activities*. Indiana University Press, Bloomington.

McDowell, L. and Court, G. (1994) 'Performing Work – Bodily Presentations in Merchant Banks', *Society and Space* 12(6): 727–50.

McLintock, A. (1995) *Imperial Leather*. Routledge, London.

Meinig, D. (1979) *The Interpretation of Ordinary Landscapes*. Oxford University Press, Oxford.

—— (1986) *The Shaping of America: A Geographical Perspective on 500 Years of History*. Yale University Press, New Haven.

Meyrowitz, J. (1985) *No Sense of Place*. Oxford University Press, Oxford.

Miller, D. (1987) *Material Culture and Mass Consumption*. Blackwell, Oxford.

Miller, R. (1991) 'Selling Mrs. Consumer: Advertising and the Creation of Suburban Socio-Spatial Relations 1910–30', *Antipode* 23(3): 263–301.

Mills, C. (1995) 'Knowledge, Gender and Empire', pp. 29–50 in Blunt, A. and Rose, G. (eds), *Writing Women and Space: Colonial and Postcolonial Geographies*. Guilford Press, New York.

—— (1996) 'Gender and Colonial Space', *Gender, Place and Culture* 3(2) 125–47.

Morley, D. and Robins, K. (1993) *Spaces of Identity*. Routledge, London.

Nag, D. (1991) 'Fashion, Gender and the Bengali Middle Class', *Public Culture* 3(2): 93–112.

Natter, W. (1993) 'The City as Cinematic Space: Modernism and Place in Berlin, Symphony of a City', pp. 203–28 in Aitken, S. and Zonn, L. (eds), *Place, Power, Situation and Spectacle: A Geography of Film*. Rowman & Litlefield, Lanham, Maryland.

Norkunas, M. (1993) *The Politics of Public Memory: Tourism, History and Ethnicity in Monterey, California*. SUNY Press, New York.

Nye, D. (1991) *Electrifying America: Social Meanings of a New Technology, 1880–1940*. MIT Press, Cambridge, MA.

O'Tuathail, G. (1997) *Critical Geopolitics*. Routledge, London.

Olsson, G (1975) *Birds in Egg, Eggs in Birds*. Pion, London.

Ong, A. (1987) *Spirits of Resistance and Capitalist Discipline: Factory Women in Malaysia*. State University of New York Press, Albany.

Parker, K. (1996) Southern Africa in Schwarz, B. (ed.), *Expanding England: Colonial Histories and Entanglements*. Routledge, London.

Parry, B. (1983) *Conrad and Imperialism: Ideological Boundaries and Visionary Frontiers*. Macmillan, London.

—— (1993) 'The Contents and Discontents of Kipling's Imperialism', in Carter, E, Donald, J. and Squires, J. (eds), *Space and Place: Theories of Identity and Location*.

Phillips, R. (1995) 'Spaces of Adventure and Cultural Politics of Masculinity', *Society and Space* 13(5): 591–608.

—— (1996a) *Masculinity and Adventure Fiction*. Guilford Press, New York.

—— (1996b) *Mapping Men and Empire: Geographies of Adventure*. Routledge, London.

Pile, S. and Thrift, N. (eds) (1996) *Mapping the Subject*. Routledge, London.

Pocock, D. (ed.) (1981) *Literature and Geography*. Croom Helm, London.

Porter, R. (1990) 'The Exotic as Erotic: Captain Cook in Tahiti', in Porter, R. and Rousseau, G. (eds), *Exoticism in the Enlightenment*. Manchester University Press, Manchester.

Porter, T. (1995) *Trust in Numbers: The Pursuit of Objectivity in Public Life*. Princeton University Press, Princeton, NJ.

Poulet, G. (1978) *Proustian Spaces*, trans. E. Coleman. Johns Hopkins University Press, Baltimore.

Pred, A. (1991) 'Spectacular Articulations of Modernity: The Stockholm Exhibition of 1897', *Geografiska Annaler* 73 B (1): 45–84.

Radhakrishnan, R. (1996) *Diasporic Mediations: Between Home and Location*. Minnesota University Press, Minneapolis.

Relph, E. (1976) *Place and Placelessness*. Pion, London.

—— (1981) *Rational Landscape and Humanistic Geography*. Croom Helm, London.

Revill, G. (1991) 'The Lark Ascending: Monument to a Radical Pastoral', *Landscape Research* 16(2): 25–30.

Richon, O. (1996) 'Representation, the Harem and the Despot', pp 242–57 in the Block Reader in *Visual Culture*. Routledge, London.

Riffenburgh, B. (1993) *The Myth of the Explorer*. Oxford University Press, Oxford.

Ritzer, G. (1993) *The McDonaldization of Society: An Investigation into the Changing Character of Contemporary Social Life*. Pine Forge Press, Thousand Oaks.

Roberts, D. (1988) 'Beyond Progress: The Museum and the Montage', *Theory, Culture & Society* 5: 543–57.

Robins, K. (1991) 'Tradition and Translation: National Culture in its Global Context', pp. 21–44 in Corner, J. and Harvey, S. (eds), *Enterprise and Heritage: Crosscurrents in National Culture*. Routledge: London.

Robinson, B. (1988) 'Literature and Everyday Life', *Antipode* 20(3): 180–206.

Rose, G. (1993) *Feminism and Geography*. Routledge, London.

Rowles, G. (1978) *Prisoners of Space? Exploring the Geographical Experience of Older People*. Westview, Boulder.

Sack, R. (1986) *Human Territoriality: Its Theory and History*. Cambridge University Press, Cambridge.

—— (1988) 'The Consumer's World: Place as Context', *Annals of the Association of American Geographers* 78(4): 642–64.

—— (1990) 'The Realm of Meaning: The Inadequacy of Human-nature Theory and the View of Mass Consumption', in Turner, B. (ed.) *The Earth as Transformed by Human Action: Global and Regional Changes in the Biosphere in the Last 300 years*. Cambridge University Press, Cambridge.

Sahlins, M. (1976) *Culture and Practical Reason*. Chicago University Press, Chicago.

Said, E. (1992) *Culture and Imperialism*. Vintage, London.

Sauer, C. (1956) 'The Education of a Geographer', *Annals of the American Association of Geographers* 46: 287–99.

—— (1962) *Land and Life: A Selection from the Writings of Carl Sauer*, ed. J. Leighley. University of California Press, Berkeley.

Schivelbusch, W. (1977) *The Railway Journey: Trains and Travel in the Nineteenth Century*. Blackwell, Oxford.

Schmid, D. (1995) 'Imagining Safe Urban Space: the Contribution of Detective Fiction to Radical Geography', *Antipode* 27(3): 242–69.

Scott, J. (1984) *Weapons of the Weak*. Yale University Press, New Haven.

—— (1990) *Domination and the Arts of Resistance*. Yale University Press, New Haven.

Seamon, D. (1980) *A Geography of the Lifeworld: Movement, Rest and Encounter*. Croom Helm, London.

Shields, R. (1989) 'Social Spatialisation and the Built Environment: The West Edmonton Mall', *Society and Space* 7: 147–64.

—— (1991) *Places on the Margin: Alternative Geographies of Modernity.* Routledge, London.

Simmel, G. (1990 [1907]) *The Philosophy of Money*. Routledge, London.

Simon, R. (1992) 'The Formal Garden in the Age of Consumer Culture: A Reading of the Twentieth Century Shopping Mall', pp. 231–50 in Franklin, W. and Steiner, M. (eds), *Mapping American Culture*. University of Iowa Press, Iowa City.

Sizemore, C. (1984) 'Reading the City as Palimpsest: The Experiential Perception of a City in Doris Lessing's *The Four-Gated City*', in Squier, S. (ed.), *Women Writers and the City*. University of Tennessee Press, Knoxsville.

Spradley, J. and Mann, B. (1975) *The Cocktail Waitress: Woman's Work in a Man's World*. Wiley, New York.

Squier, S.M. (ed.) (1984) *Women Writers and the City*. University of Tennessee Press, Knoxville.

Squire, S. (1988) 'Wordsworth and Lake District Tourism: Romantic Reshaping of the Landscape', *Canadian Geographer* 32(3): 237–47.

—— (1990) 'Wordsworth and Lake District Tourism: A Reply', *Canadian Geographer* 32(2): 164–70.

—— (1993) 'Valuing the Countryside: Reflections on Beatrix Potter Tourism', *Area* 25(1): 5–10.

—— (1994) 'The Cultural Values of Literary Tourism', *Annals of Tourism Research* 21(1): 103–20.

Stallybrass, P. and White, A. (1986) *The Politics and Poetics of Transgression*. Methuen, London.

Stocking, G. (ed.) (1974) *The Shaping of American Anthropology, 1883–1911*. Basic Books, New York.

Taussig, M. (1980) *The Devil and Commodity Fetishism*. University of North Carolina Press, Chapel Hill.

Tester, K. (ed.) (1995) *The Flâneur*. Routledge, London.

Thompson, E. (1962) *The Making of the English Working Class*. Pelican Books, London.

Thornton, S. (1995) *Club Cultures*. Routledge, London.

Thrift, N. (1981) 'Literature, the Production of Culture and the Politics of Place', *Antipode* 12: 12–23.

—— (1995) 'Speed Light and Power', in Cloke, P. (ed.), *Writing the Rural*. Paul Chapman, London.

Thrift, N. and Glennie, P. (1993) 'Historical Geographies of Urban Life and Modern Consumption', pp. 33–48 in Philo, C. and Kearns, G. (eds), *Selling Places: The City as Cultural Capital, Past and Present*. Pergamon Press, Oxford.

Tuan, Y. (1992) 'Place and Culture: Analeptic for Individuality and the World's Indifference', pp. 27–50 in Franklin, W. and Steiner, M. (eds), *Mapping American Culture*. University of Iowa Press, Iowa City.

Venturi, R. (1973) *Learning from Las Vegas*. Harvard University Press, Cambridge, MA.

Warren, B. (1986) 'Citizens of Empire: Baden-Powell, Scouts and Guides as an Imperial Ideal', in Mackenzie, J. (ed.) *Imperialism and Popular Culture*.

Williams, R. (1973) *The Country and the City*. Chatto & Windus, London.

—— (1977) *Marxism and Literature*. Blackwell, Oxford.

—— (1987) *Television: Technology and Cultural Form*. Routledge, London.

—— (1985) *Dreamworlds of Desire: Mass Consumption in Late Nineteenth-Century France*. University of California Press, Berkeley.

Williamson, T. (1995) *Polite Landscapes: Garden and Society in Eighteenth-Century England*. Johns Hopkins University Press, Baltimore.

Williamson, B. (1982) *Class, Culture and Community: A Biographical Study of Social Change in Mining*. Routledge, London.

Willis, S. (1991) *A Primer for Daily Life*. Routledge, London.

Wilson, E. (1991) *The Sphinx in the City*. Virago, London.

Wooden, W. (1995) *Renegade Kids, Suburban Outlaws: From Youth Culture to Delinquency*. Wadsworth, Belmont.

Woodward, R (1993) 'One Place, Two Stories: Two Interpretations of Spitalfields in the Debate over its Redevelopment', pp. 253–266 in Philo, C. and Kearns, G. (eds) *Selling Places: The City as Cultural Capital, Past and Present*. Pergamon Press, Oxford.

Wright, P. (1985) *On Living in an Old Country: The National Past in Contemporary Britain*. Verso, London.

Zelinsky, W. (1973) *The Cultural Geography of America*. Princeton University Press, Princeton, NJ.

Zukin, S. (1982) *Loft Living: Culture and Capital in Urban Change*. Radius, London.

Zukin, S. *et al.* (1992) 'Bubbling Cauldron: Global and Local Interactions in New York City Restaurants, in Smith, M. (ed.), *After Modernism: Global Restructuring and the Changing Boundaries of City Life*. Transaction, New Brunswick.

—— (1995) 'Bubbling Cauldron', revised version of Zukin *et al.* (1992), in Zukin, S. *The Cultures of Cities*. Blackwell, Oxford.

Index

Note: page references in **bold** refer to
chapters or glossary definitions

CPSIA information can be obtained
at www.ICGtesting.com
Printed in the USA
LVHW08s0211260818
588076LV00001B/3/P